From Safety to Safely

The conventional interpretation of safety, known as Safety-I, denotes a condition where as little as possible goes wrong, and the focus of practical efforts in management or analysis is on the occurrence of unacceptable outcomes and on how to reduce their number to an acceptable level, ideally zero. The emphasis is therefore on how to manage safety as such, as seen in the ubiquitous safety management systems (SMS). As Professor James Reason astutely points out, this raises the interesting question of how it is possible to learn about something, let alone manage it, if it is studied only in situations in which it is absent. The solution proposed by and described in this book is to stop using safety as a noun and instead to use it as an adverb: safely.

Now often referred to as Safety-II, this solution is the logical consequence of resilience engineering and will require new methods, several of which already exist and have proved their worth in practice for years. The question ceases being what to manage and becomes how to manage. Managing safety is protective, hence a non-productive cost, which at best avoids a loss. Conversely, managing safely is productive and can generate revenue in addition to preventing or avoiding losses; aviation and mining are prime examples.

From Safety to Safely provides a practical perspective on managing safely, illustrating a practical form of synesis. It offers a new understanding of safety, combining concerns for productivity and safety rather than juxtaposing them, and it shows how to manage complex industrial and social systems in the spirit of resilience engineering and synesis. It is the first book to completely dispense with the loaded term "safety" while offering a practical and viable alternative. Spoiler alert: this book does not mention or analyse any celebrated accidents.

This book is for all middle and senior managers, board members, and independent consultants seeking to ensure safe, revenue-generating operations.

Erik Hollnagel is Scientific Director at the Institute of Resilient Systems+, Seoul, South Korea; Visiting Professorial Fellow, Faculty of Medicine and Health Sciences at Macquarie University, Australia; Visiting Fellow of the Institute for Advanced Study at the Technische Universitat München, Germany; and Professor Emeritus at universities in Sweden, France, and Denmark. His work focuses on unified system change and management. Erik is the author of more than 500 publications including articles from recognised journals, conference papers, and reports as well as 28 books, and he is still struggling to make sense of the blooming, buzzing confusion.

"Even deeply-read safety scholars will profitably immerse themselves in this latest book from Professor Hollnagel. Not just scholarly readers, too. Everyone – researchers, business people, safety practitioners for example – will learn from engaging in this splendid book."

Professor Jeffrey Braithwaite,
Professor and Director,
Australian Institute of Health Innovation,
Macquarie University, Australia

"After acronyms and atavism, Erik is now resorting to a clever grammatical twist to inspire the reformation instead. He's deploying just one letter substitution. And taking us from noun to adverb. From safety to safely. The shift this heralds, however, is not subtle at all. In fact, it's huge. We've been managing safety for its absence. Which, if you think about it (which Erik has done a lot), is not only illogical but profoundly stupid. "Managing the primary process of a system or a company well is after all what provides the basis for productivity and business, regardless of domain and type of activity," Erik writes. Managing it safely means doing just that, and assuring that you can keep doing it – as long as you learn from what goes well, and why, and then commit to doing ever more of it. Who can disagree with that?"

Professor Sidney Dekker,
Griffith University, Australia

"This new book is a profound exploration of the prevailing interpretation of safety, taking us on a captivating journey through the foundations and assumptions underpinning our approach. It's a groundbreaking work which prompts us to reconsider the very essence of safety management."

Professor Dr Andrew Sharman,
Managing Director, RMS Switzerland, Switzerland

From Safety to Safely

Principles and Practice of Systemic
Potentials Management

Erik Hollnagel

Routledge
Taylor & Francis Group

LONDON AND NEW YORK

Designed cover image: Erik Hollnagel

First published 2026
by Routledge
4 Park Square, Milton Park, Abingdon, Oxon OX14 4RN

and by Routledge
605 Third Avenue, New York, NY 10158

Routledge is an imprint of the Taylor & Francis Group, an informa business

British Library Cataloguing-in-Publication Data
A catalogue record for this book is available from the British Library

Library of Congress Cataloging-in-Publication Data
Names: Hollnagel, Erik, 1941– author.
Title: From safety to safely : principles and practice of systemic
    potentials management / Erik Hollnagel.
Description: Abingdon, Oxon ; New York, NY : Routledge, 2025. |
    Includes bibliographical references and index.
Identifiers: LCCN 2024027961 (print) | LCCN 2024027962 (ebook) |
    ISBN 9781032664712 (hardback) | ISBN 9781032664705 (paperback) |
    ISBN 9781032664729 (ebook)
Subjects: LCSH: Industrial safety.
Classification: LCC T55 .H6525 2025 (print) | LCC T55 (ebook) |
    DDC 363.11—dc23/eng/20241108
LC record available at https://lccn.loc.gov/2024027961
LC ebook record available at https://lccn.loc.gov/2024027962

ISBN: 9781032664712 (hbk)
ISBN: 9781032664705 (pbk)
ISBN: 9781032664729 (ebk)

DOI: 10.4324/9781032664729

Typeset in Bembo Std
by Apex CoVantage, LLC

This book is dedicated to Andrew Hale.

Contents

List of figures

List of tables

Foreword

"Reformers tend to be difficult people," Michael Kinsley of Vanity Fair wrote in 2014, "but they come in different flavors." I quoted Kinsley in a review of one of Erik's previous books (*Safety I and Safety II*) (Dekker, 2014), and the quote is as apt now as it was then. There were Robespierre, Trotsky – revolutionaries who declared the ancien régime corrupt and called for it to be toppled. Then there were the ascetics, who proclaimed to know or understand less of the foil they had chosen to challenge with each day that went by yet still contentedly racked up results while false-modestly rebuffing praise or adulation.

Erik Hollnagel is a reformer, I wrote in 2014, if ever we had one in our field. I believe that to be true today, and particularly – again – with this latest book. There are probably enough people who would cast Erik comfortably in any of Kinsley's roles. I would imagine him most confidently somewhere between revolutionary and ascetic. And not necessarily more difficult than the next reformer. Among the targets that have quite deservingly made it into Erik's sight over the past decades are information processing psychology, human reliability analysis, and the very notion of "human error" – to name but a few. He has even taken the practice of accident investigation to task for being the enemy of learning from failure – which we get back in a nice dose in this book too.

This time around, though, the ancien régime that needs to fold – and at whose insufferable follies he shakes his head – is wrapped up most efficiently or mendaciously in a label that has superciliously committed but underinformed or misguided management types joining cheerleader ranks: Vision Zero. When you reach Erik's stature, it is neither illegitimate nor costly to be a tad irascible – which he's always had in him but which can now sardonically bubble forth from his prose with impunity. The logic of Vision Zero (which maintains that safety equals that as little as possible, ideally nothing, goes wrong) "leads to extensive and expensive accident investigations that are of limited value, since eliminating failures to prevent that something goes wrong does not as such contribute to work going well."

Boom.

Happy days, accident investigators . . . or those beleaguered safety minions who have scores of dubiously relevant incident investigations languishing in their SMS which "haven't been 'closed-out' yet" (whatever *that* means). Erik would probably shrug, shake his head, walk away, and go do something more useful. His rightful disdain of SMS is in plain sight in any case. So next time the regulator comes to visit to whine about the SMS not being up to date (as in: too many investigations that haven't been "closed out" yet), tell them to take Erik out for a coffee.

Erik had his fun with cute, memorable acronyms back in the 1990s and 2000s. Think of CREAM, or FRAM, or ETTO. As his wit ripened ever further over the years, he plumbed abstruse and ancient terms (I mean, who had ever heard of *synesis* before Erik?) or even co-opted Latin to make a contrasting point. For example, in this book he talks about a logical alternative to "Vision Zero" (or perhaps, playing a little more with Latin and some Greek, he might have called it "inverse" or "converse" or "obverse" or "antithesis"). This is "Visio Centum" (Latin for one hundred), meaning that 100 percent of work should go well. Those familiar with Safety II, in which we aim for as much as possible to go well (as opposed to nothing going wrong), may give *Visio Centum* a half-smiling, insider shrug. Those committed to Zero will have no idea what Erik is on about. Maybe they won't even read this book. Which would be a waste, because after acronyms and atavism, Erik is now resorting to a clever grammatical twist to inspire the reformation instead. More subtle, less CAPITAL-letters-in-your-face. Indeed, quite minimalist. He's deploying just one-letter substitution. And taking us from noun to adverb. From safety to safely.

The shift this heralds, however, is not subtle at all. In fact, it's huge. We've been managing safety for its absence. Which, if you think about it (which Erik has done a lot), is not only illogical but profoundly stupid. Perhaps this is not such a huge insight, then, because it is also utterly common-sensical. "Managing the primary process of a system or a company well is after all what provides the basis for productivity and business, regardless of domain and type of activity," Erik writes. Managing it safely means doing just that and assuring that you can keep doing it – as long as you learn from what goes well, and why, and then commit to doing ever more of it. Who can disagree with that?

Not that this is necessarily straightforward. It will cost a bit of effort, because, as Erik reminds us (and spends a good portion of the book unpacking under the label of "complexity"), "successful outcomes are known to emerge from an intricate and fortuitous combination of factors, conditions and circumstances, that we are unable to predict." Some would ask whether we can manage any-thing safely if we're unable to predict the fortuitous combination of factors, conditions, and circumstances that make it so. But there is. As Erik says,

> In the change from safety to safely it is no longer safety, but rather the pri-mary process of a system or a company that is managed, and managed safely, so that as much as possible goes well. There is therefore something tangible to manage, be it the production of some physical artefact or the provision of some kind of service.

Do what you do, but do it even better, and try to understand where success in doing so comes from. Erik even has a term for this (too): systemic potentials management. You'll find it in Part IV of this book.

From our legacy and cravings for certainty – much of it no good to the future of safety – through complexity, the futility of accident investigation, and the management and assessment of systemic potentials that allow things to go well, Erik ends with sceptical expressions about our ability to divine the future by great thinkers (thankfully not *all* of them men), perhaps in an effort to pull down expectations of our ability to manage that future – even with the "systemic potential" of this little book in hand.

Erik brings with him a wit and intellectual horsepower that few of us are adequately equipped to handle. He has always been indefatigable. Though if – and when – he ever tires of writing a particular book (which all impatient intellectuals do: finishing one is never as exciting as starting one, which in turn is not as exciting as the process of thinking about the one you're going to write next), we are all winners, because we end up with characteristically and mercifully short and thin books from Erik.

Sidney Dekker
Brisbane, 2024

Reference

Dekker, S. W. A. (2014). Safety I and Safety II. *Journal of Contingencies and Crisis Management*, *22*(4), 239–240.

Foreword

For the past 26 years I've worked in occupational safety and health (OSH). I've picked severed body parts out from within machinery, explained to paramedics the cause of injuries, told loved ones that their respective other isn't coming home, and watched families fall apart. I've filled out more forms, spreadsheets, and templates than I care to remember. I've coached leaders on how to talk about safety and been asked to leave when they didn't like what I said. I've stood on stage in front of hundreds of people anticipating something useful and had CEOs cry on my shoulder. I've worked with leaders in more than 130 countries around the world and read books like they're going out of fashion – including most of the 29 books that the author of this one has written. A constant over this time has been my admiration of the thinking of Professor Dr Erik Hollnagel. It's therefore an immense privilege to write some words here to welcome you to this, his latest book.

During my 26 years in safety (and before that, too) there has been much work undertaken on the wide (and ever-growing) notion of "workplace safety." Without doubt, some of these works – journal articles, conference papers, or books – have been of significant importance, both practically and from a perspective of broadening thought. Most often though, efforts seem to fall into one of two categories.

The first is the dark morass where overly-technical, jargonised pages – incomprehensible to most of us – lurk. Out of sight, out of mind. Over the years I've ventured many, many times to this place, seeking wisdom – sometimes finding it, most often rather returning, covered in metaphorical mud, wondering how so much deep thought could be so lost to what I imagine must be its true goal: practical application for the benefit of all.

The second reveals a tendency for authors to proclaim that their way of "doing safety" – their model, philosophy, position, or perspective – is *The Right One*. As a result, rather than creating a collaborative approach to solving the perennial problem of people getting injured, becoming sick, or losing their lives at work, these efforts that purport to providing the silver bullet, the solution, the *best* way, have generated silos. Should we focus on systems or culture? Take a risk-based approach or improve leadership? Should we do HRA, or HOP, skip, SCRUM and be AGILE? Should we get back to basics or do safety differently? Prevent accidents or create safety? Should we be "ultra-adaptive" or "high reliability"?

With the greatest of respect, I'd venture that Erik's previous books may be experienced by some readers to be, at times, in one of the two aforementioned categories. The book you now have in your hands falls into neither.

The prevailing view for many people around the world is that "safety" must mean the absence of unwanted outcomes, such as incidents or accidents. As a result, focus on "safety improvement" tends to follow a pattern of identifying failures and malfunctions of technology, procedures, workers, and their organisations and striving to limit the number of things that can go wrong. This traditional approach, (since Hollnagel's early contributions to resources such as the Resilient Health Care Net[1]) has been referred to as "Safety-I." Often associated with the traditional view of safety, Safety-I focusses on the prevention of accidents and incidents. It operates on the assumption that things go wrong because of the presence of negative factors, and the goal is to reduce or eliminate these factors to ensure safety.

Despite having been the predominant approach – rapidly spreading through the 1960–1990s in safety critical sectors such as nuclear power, aviation, etc. – we might begin to doubt whether this approach is sufficient to resolve the very real and growing problem of workplace accidents, ill-health, injuries, harm, and death.

As a response to the slow slide of safety statistics, Hollnagel offered a thoughtful, considered challenge with the suggestion of Safety-II. Defined as a practice to "look for what goes right, to focus on frequent events, to maintain a sensitivity to the possibility of failure, to wisely balance there in the sufficiency, and to view an investment and safety as an investment in productivity," Safety-II represents a more proactive, adaptive approach to safety. It acknowledges that safety is not just the absence of failures but also involves understanding and recognising that in complex systems, things often go right due to the presence of positive performance, rather than solely focusing on what goes wrong. Safety-II emphasises understanding how work is normally done successfully, as well as how people adapt to respond to varying conditions.

Safety-II, without doubt, immediately helped "safety" start to feel sexier, more alive, more relevant, more business-like. The world had changed, and now, so too, had safety. One small step for this man, one giant leap for (hu)mankind.

"I've got some regrets . . ." says the man sitting in the deep leather armchair – the colour of fine Bordeaux wine – in the Business Lounge of Stockholm's Arlanda airport. I know my friend Erik Hollnagel to be a deeply reflective man, so I'm not surprised to hear this from him. "I know I need to do something about this" he continues, "but I'm almost wishing I hadn't come up with the idea in the first place."

We'd just been running a two-day workshop for an association of Swedish organisations, where, the previous day over a meal together, he and I had been debating modern approaches to workplace safety improvement. Directly before the meal, Erik had been holding forth with the audience on the topics of performance variability, transient phenomena and emergence (rather than causality), and the manifestations of things that go well. He'd set out his argument so well that I could see heads nodding and hands furiously scribbling notes. As the applause continued, Erik calmly left the stage and I noticed it for the first time. That look on his face. As I stepped up to share my views, the predominant thought in my head was a nagging one: "Is Erik okay?"

Erik Hollnagel takes his work very seriously. Even with the loudest, longest applause his face reveals very little. Impassive, inexpressive, inscrutable. Not disinterest or deadpan, more . . . just *quiet*. I guess Edgar Schein would have said this was Erik's own sense of humility.

But as he exited the stage, just for a moment – just a second or two – something changed; Erik didn't seem happy. I noticed it, though I doubt anyone else had. Whatever it was had escaped the public view. But now, with a glass of water in hand, calmly installed in the airport lounge, he began to explain. . . .

Erik calmly began to dissect his conundrum. In the creation of Safety-II, people had jumped on this "new way of thinking" and adopted it with gusto. Erik lamented this enthusiasm, sad at the way people had misunderstood that his ideas for Safety-II were really just a build on what he'd previously referred to as Safety-I. He had never intended to suggest a "new model" that replaced what went before.

I remind him of one of his countrymen, Søren Kierkegaard, who wrote: "Livet forstås baglæns men må leves forlæns." In English it becomes "Life is experienced backwards but must be lived forwards" and I suggest it to be a resonant reminder of the dual importance of experimentation and active learning.

In a White Paper written in 2015, Hollnagel talked of the "growing interest in improving . . . safety" and lamented that "Despite decades of attention, activity and investment, improvement has been glacially slow." He wasn't wrong. And still isn't. Around the world today, safety appears fixed on boardroom agendas yet at the same time the International Labour Organisation recently revised their estimation of work-related deaths due to accident and ill-health from 2.78 million to now over 2.9 million. We might need to face the facts: the way we're working isn't working.

And so, reflecting on his previous experience Erik puts pen to paper again in a new attempt to live forward.

True to his words in his previous book,

> It may also happen that the very concept of safety is gradually dissolved, at least in the way that it's used currently, as something distinctly different from, e.g., quality, productivity, efficiency, etc. If that happens – and several signs seem to indicate that it will – then the result will not be a Safety-III, but rather a whole new concept.[2]

Hollnagel even gives a hint at what this "whole new concept" might be when he suggests (perhaps leveraging the thoughts of Karl Weick) that "safety *is a dynamic event, and something that must be created constantly and continuously* (emphasis added)."

In the ever-evolving landscape of safety and resilience engineering, Professor Erik Hollnagel has himself created constantly and continuously – challenging conventional wisdom and prevailing norms and offering groundbreaking insight. This new book, then, is a profound exploration of the prevailing interpretation of safety, taking us on a captivating journey through the foundations and assumptions underpinning our approach.

It's a groundbreaking work which prompts us to reconsider the very essence of safety management. As we embark on this intellectual journey with Professor Hollnagel, let us be prepared to confront the uncomfortable truths and embrace a new vision, one where safety transcends a mere absence of failure, leading us towards a future where work goes well, not just incident free. From Safety, to Safely.

Professor Dr Andrew Sharman
Geneva, Switzerland, February 2024

Notes

1 see www.ResilientHealthCare.net
2 Hollnagel, E. 2014. Safety-I and Safety-II, p. 178.

I   The safety legacy

Legacy prologue

According to the conventional interpretation of safety, now often called Safety-I (Hollnagel, 2014b), safety refers to a condition where as little as possible goes wrong. The American Society of Safety Engineers and the American National Standards Institute, for instance, define safety as "the freedom from unacceptable risk" (ASSE, 2011, p. 13). The focus of practical efforts whether in management, design, or analysis is consequently on the occurrence of unacceptable outcomes and on finding the means to reduce these to an acceptable level, ideally zero, as in zero fatal accidents. The emphasis is thus on how to manage safety as such, as seen by the ubiquitous safety management systems (SMS). Similarly, the International Civil Aviation Organization (ICAO, 2013) defines safety as "the state in which the possibility of harm to persons or of property damage is reduced to, and maintained at or below, an acceptable level through a continuing process of hazard identification and safety risk management" (ICAO, 2013, pp. 1–2). The ASSE and ICAO both use the relative terms *acceptable* and *unacceptable* even though they are not scientifically precise. In practice they usually mean *affordable* and *unaffordable* and are therefore relative rather than absolute.

Part I will provide a characterisation of the current situation with regard to safety and safety management where the universally accepted goal is to have zero accidents and incidents (also known as vision zero Zwetsloot et al., 2013) based on the common but mistaken assumption that this can be achieved by finding the root causes of accidents and eliminating said causes. This goal reflects attitudes and assumptions from the early 1930s which unsurprisingly are better suited to work environments and conditions 90 years ago than to the work environments and conditions existing today. Part I also develops the logical alternative to vision zero, which is *visio centum* (Latin for "one hundred"), meaning that 100 per cent of work should go well. The origin of vision zero is identified, and it can be argued that vision zero leads to extensive and expensive accident investigations (Morris et al., 1998) that are of limited value, since eliminating failures to prevent that something goes wrong does not as such contribute to work going well. Following that, the perspectives of Safety-I and Safety-II are summarised and contrasted. Part I concludes with some comments on how the

DOI: 10.4324/9781032664729-1

human craving for certainty often stands in the way of a proper understanding of how things actually work and happen, precisely captured by the statement by the American psychoanalyst Erich Fromm: "The quest for certainty blocks the search for meaning."

Since Safety-I focuses on what does not go well, it is reasonable to begin by listing the implicit assumptions behind this view of the world. Because they are implicit, they are tacitly accepted throughout the field and therefore rarely if ever scrutinised or questioned!

Vision zero and visio centum

Safety-I and Vision zero have a common conceptual basis or logic or a set of general reasons why accidents happen. These have been summarised in the left-hand column of Table IV.1. Yet even the most pessimistic safety manager must admit that not every action goes wrong or fails, but that a fair number of them, in fact the vast majority, actually go well. Otherwise he and the company he works for would soon be out of business. There must therefore also be a set of general reasons why accidents do not always happen. These have also been summarised in Table I.1.

The logic of vision zero

The default reasoning of vision zero goes as follows:

1. The overreaching goal for any system or company is to avoid accidents.
2. The only way to ensure that there will be no accidents is to make sure that nothing fails. This can best be done by finding and eliminating the probable causes of failures!!!
3. It follows from the previous point that humans (as a recognised source of failure) are a liability and that performance variability (whether in the shape of violations or deviations) is a threat.
4. The purpose of design is to constrain variability, in order to prevent adverse outcomes.

Table I.1 How the world works according to vision zero or Safety-I

Accidents happen because	"Nothing" happens because
Components (HW/SW) will fail sooner or later	Systems are well designed and scrupulously maintained
There will always be unexpected and unrecognised situations	Procedures are complete and correct
Combinations of components can hide sneak faults and other flaws	Designers can anticipate and prepare for every contingency

5. And since work usually goes well (Amalberti, 2001), there is no real need to pay attention to it.
6. Instead we should worry about why work sometimes goes wrong.

The regulator's paradox

Zero Accident Vision is a philosophy which states that nobody should be injured due to an accident. It is more a way of thinking and an ethical rather than a numerical goal. In terms of accident prevention strategies, Zero Accident Vision can be viewed as a way of thinking, which claims that all accidents can be prevented by finding and eliminating the causes when something has gone wrong. But this unfortunately also eliminates the basis for learning and therefore invokes the regulator's paradox as described by (Weinberg & Weinberg, 1979).

> The task of a regulator is to eliminate variation, but this variation is the ultimate source of information about the quality of its work. Therefore, the better the job a regulator does the less information it gets about how to improve.
>
> (Weinberg & Weinberg, 1979, p. 250)

The regulator's paradox means that vision zero, unlike visio centum, is self-defeating. The purpose of Safety-I (Hollnagel, 2014b) is widely accepted to be that there are no – or zero – accidents at work. In this view safety is a static outcome; it is something that exists as a result of something that has happened in the past but does not happen any longer and safety is therefore grammatically used as a *noun*.

Note that there are some obvious contradictions and inconsistencies in the two sets of assumptions in all three rows.

Managing safety, the logic of vision zero

The idea of managing safety is thus central to vision zero. Whether this is possible depends critically on the definition of safety, of which a number are provided in the following section.

Assorted definitions of safety

Safety according to the ICAO

According to the International Civil Aviation Organization (ICAO), "Safety is the state in which the risk of harm to persons or of property damage is reduced to, and maintained at or below, an acceptable level through a continuing process of hazard identification and risk management" (ICAO, 2013, pp. 1–2).

Safety according to the AHRQ

The Agency for Healthcare Research and Quality (AHRQ) defines safety as "freedom from accidental injury," which can be achieved by "avoiding injuries or harm to patients from care that is intended to help them."

A pragmatic industry definition of safety

The French energy company TOTAL has a more pragmatic definition, which in many ways is similar to the definition of resilience:

> Industrial safety can be defined as the ability to manage the risks inherent to operations or related to the environment. Industrial safety is not a dislike of risks; rather it is a commitment to clearly identify them in relation to production operations, assess them in terms of quality and quantity, and manage them.
>
> (Jaubert, 2006)

Safety according to the WHO

It is interesting to compare the previous definitions with how the World Health Organization (WHO) looks at health. The analogy between safety and health should not require any elaboration.

The World Health Organization defines health as "a state of complete physical, mental, and social well-being and not merely the absence of disease or infirmity." Safety would in this manner be defined as *more than* the absence of accidents and incidents.

In the same way, Mosby's Medical Dictionary (2009) defines physical fitness as "the ability to carry out daily tasks with alertness and vigour, without undue fatigue, and with enough energy reserve to meet emergencies or to enjoy leisure time pursuits," which not coincidently is very similar to the definition of resilience as the ability of a system to function as required under expected and unexpected conditions alike as required. A second definition of physical fitness is "the state of being suitably adapted to an environment" (American Heritage Stedman's Medical Dictionary, 2008).

Newer definitions of safety

Common to the previous definitions is that safety is defined by its absence – WHO's definition excepted, but then this was not intended as a definition of safety as such. The previous definitions of safety differ from definitions based on a modern systemic perspective, for instance these:

Safety as a dynamic non-event

"Reliability is a dynamic non-event . . . it is an ongoing condition in which problems are momentarily under control due to compensating changes" Weick (1987, p. 118).

This definition was nominally about reliability rather than safety, but if we allow the two terms to be substituted it reads:

> Safety is a dynamic non-event . . . it is an ongoing condition in which problems are momentarily under control due to compensating changes.

Which really is another way of saying that work goes well. Accidents are the events we pay attention to. Work that goes well in between the occasional accidents are the dynamic non-events, hence the absence of accidents but the presence of safety, and we usually do not pay attention to them, not least because they happen all the time.

Safety as absent or as present?

> Safety is defined and measured more by its absence than by its presence.
> Reason (2000, p. 3).

This seemingly paradoxical statement refers to the fact that immediately after an accident people are prone to say that it was due to the absence or lack of safety or that safety was missing or deficient.

Is vision zero really possible?

In summary, vision zero is not realistically possible. There are two reasons for that. One is the root cause fallacy described in the following section. The other is that the solution of eliminating the root causes, provided it actually exists, only is effective if it can be assumed that the same conditions or situations will recur and that there never will be new situations or conditions. But that is impossible given the complexity (see Part II) of the socio-technical systems we have to cope with. For these reasons alone it is impossible to manage safety. The alternative proposed here is to manage safely in accordance with the visio centum.

The safety mantra

A mantra is a sacred verbal formula repeated in prayer, meditation, or incantation, such as an invocation of a god, a magic spell, or a syllable or portion of scripture containing mystical potentialities. The typical statements issued by responsible managers or authorities after a serious accident or unexpected events are often so similar that they appear like a sacred rite, to appease either the gods or the general public. Here are two examples of that:

Bulgarian bus crash incident

On November 23, 2021 a bus carrying North Macedonian tourists crashed when it hit a highway barrier and burst into flames. At least 45 people, including

12 children, died in the flames. When Bulgaria's interim Prime Minister Stefan Yanev visited the crash site he described the incident as "an enormous tragedy" and continued, "let's hope we learn lessons from this tragic incident and we can prevent such incidents in the future." This may sound familiar because similar statements are common in response to an accident. In this case it was quite appropriate, but that is not always the case.

United Airlines accident

On March 14, 2018, something unexpected happened on a United Airlines (UA) flight from Houston to New York. After the event the airline issued the following statement

> This was a tragic accident that should never have happened. We apologise to everyone impacted by this breach and are working hard to ensure it does not happen again.
>
> (BBC)

United Airlines has accepted "full responsibility." What had happened was that a flight attendant (mistakenly) had told a family to place their French bulldog in the overhead locker where it suffocated en route after having barked for a couple of hours. I do not mean to belittle this incident, which certainly affected the family much. But I find it interesting that the official reaction is so eerily similar to the reaction after the bush crash where 57 humans died in the burning bus. (The bus crash, where 45 humans burned to death, was called an incident, whereas this event, where a little dog suffocated, was called an accident!) The similarity in the statements strongly suggests that the soothing words serve more as a ritual or a mantra than as a carefully considered response to the factual event. A generic safety mantra would basically go like this:

1. This was a tragic accident that should never have happened, and we are very sorry that it did.
2. We will do our utmost to learn the necessary lessons to make sure that something like this never happens again. (The last part is strictly speaking not necessary, because it could only happen again if accidents are the inevitable outcome of an immutable sequence of events that occurs in exactly the same way a second time. But unwanted outcomes are the result of a unique combination of events and conditions as the philosopher Bertrand Russell pointed out many years ago (the quotation is reproduced in the description of the Heinrich dogma that follows.))

It is interesting to compare the safety mantra to the paraphrased statement from the fictitious Capt. Jean-Luc Picard, of the Starship Enterprise: "it is vital to anticipate, in order not to err in the same way even once."

That would be a worthy ambition for both safety management and managing safely.

The safety mantra is not the prerogative of the "typical" industrial accidents. During Christmas dinner at the company Airbus Atlantic in 2023, more than 100 persons became ill from suspected food poisoning. Predictably the company vowed to "to identify the cause of the illness and ensure this cannot happen again in the future." Perhaps the safety mantra is actually part of the curriculum of MBA courses?

The Heinrich dogma

The Heinrich dogma was introduced in the first published book on industrial accident prevention (Heinrich, 1931, p. 39). The dogma states that "It is widely accepted as true that the cure of a given troublesome condition depends primarily upon knowledge of its cause and the elimination, or at least the mitigation, of that cause." But one should always be careful with widely accepted truths. The Heinrich dogma actually hides three key assumptions:

- *First* that an outcome always can be attributed to a specific cause, (actually the law of causality in reverse)
- *Second* that it is possible to find or determine what that cause was or is (provided enough data is gathered and enough time and efforts are spent.
- *Third* that the elimination of the cause effectively will prevent the outcome from occurring again (the law of causality one more time). The dogma thus rests on the principles of simple linear causality, specifically that it is possible to reason backwards from final effect to the initial or root cause (see comments about the root cause fallacy in the following discussion.) While this may have been widely and uncritically accepted in 1931 and possibly even adequate for the work environment and the nature of work that existed then, it is by no means reasonable or acceptable today. It is hardly necessary to point out that the industrial work environments of 1931 bear little resemblance to today's complex socio-technical systems.

The Heinrich dogma represents the common belief that the best and, indeed, the only way to prevent accidents, is to learn from them by studying them – especially to find the causes and then eliminate or neutralise these (Kletz, 2001). But the law of causality is flawed as the eminent philosopher Bertrand Russell argued, although it has gone unnoticed because it was not written in relation to safety:

> The law of causality, I believe, like much that passes muster among philosophers, is a relic of a bygone age, surviving, like the monarchy, only because it is erroneously supposed to do no harm. . . . The principle, same cause, same effect, which philosophers imagine be vital to science is therefore

utterly otiose. As soon as the antecedents have been given fully to enable the consequent to be calculated with some exactitude, the antecedents have become so complicated that it is very unlikely they will ever recur. Hence, if this were the principle involved, science would remain utterly sterile. . . . No doubt the reason why the old "law of causality" has so long continued to pervade the books of philosophers is simply that the idea of a function is unfamiliar to most of them, and therefore they seek an unduly simplified statement. There is no question of repetitions of the "same" cause produc-ing the "same" effect: it is not in any sameness of causes and effects that the constancy of scientific laws consists, but in the sameness of relations. And even "sameness of relations" is too simple a phrase.

(Russell, 1913)

The root cause fallacy

Even though Heinrich (1931) did not use the actual term root cause, he did refer to the first domino falling as the root of the trouble (p. 106), which essentially is the same thing. It also corresponds to the central statement of the dogma that: "the cure of a given troublesome condition depends primarily upon knowledge of its cause and the elimination, or at least the mitigation, of that cause." The idea of a first cause can arguably be traced back to Aristotle's theory of causa-tion, which distinguished among four causes: material, formal, efficient, and final but did not include a root or "first" cause as such, yet the "root" of this way of thinking lies even further back. Every culture and religion includes an account of how the world began. The questions "where do we came from" and "why we are here" are also something that science and philosophy have been addressing. Two examples will suffice.

Darwin's theory of evolution

While the theory of evolution, still disputed to this very day, does not explain the reason why we are here, it at least provides an explanation for why we are as we are, the physiological – and perhaps also the psychological – characteristics of homo sapiens. It also suggests it is possible to reason backward to the supposed origin, known as the Last Universal Common Ancestor or (LUCA), which therefore essentially is at the root of the species we know today. In this case there is, however, an articulated theory that supports and justifies reasoning backward and also empirical evidence to support the theory. It is remarkable that Darwin in his notebook drew a sketch of the tree of life which seductively suggested that it was possible to go back to the root, except the root of a tree is part of a complicated mycorrhizal network (Simard, 1921); the trunk of a tree does not continue below ground like a pole or a carrot, as anyone who has ever tended a garden will know. Trees actually are far more complicated and intertwined below ground than above ground.

The big bang

Another example from science is the famous big bang theory. This is also an evolutionary theory, although on a somewhat larger scale, as it tries to describe how the universe about 13,000 billion years ago suddenly expanded from an initial state of high density and temperature (called the singularity). This is an excellent example of how it is possible to reason backward following the laws of physics until the beginning of time, or at least to 10^{-6} seconds after time began. The outstanding question is, of course, what started it all, where life came from, and why the universe suddenly about 13,000 billion years ago went from a stable state to starting to expand, an expansion that continues to this very day. Unlike Darwin's theory of evolution, it is possible to confirm the big bang theory by empirical evidence, and even better it is possible to make predictions that can be tested and confirmed, such as the cosmic microwave background (CMB) radiation. But when it comes to what started it all we are no wiser than with Darwinian evolution, at least if the intervention of higher powers is ruled out. In the case of evolution two contenders are the primordial soup hypothesis, supported by the Miller–Urey experiment in 1953 and the panspermia hypothesis (the hypothesis on the interstellar spreading of primordial life), which can be traced back to Anaxagoras (c. 500–c. 428 BCE) but which otherwise is not supported by anything. The panspermia hypothesis actually cheats by placing the origin of life somewhere else. Physics and cosmology, on the other hand, have a familiar kind of solution, namely to propose hitherto unknown factors to explain what happened. In the case of the Big Bang theory, it is the convenient and timely appearance of an inflation field – or even a hypothetical particle called an inflation, neither of which, of course, exists any longer! This is not too different from what safety science and human factors do, when they propose hypothetical causes and even counterfactual conditions (human error, situation awareness, and safety culture to name but three). But physics is on firmer ground, as proven by the case of the Higgs' boson. Dark matter and dark energy are a different story and are perhaps more like safety culture; "human error" can be considered the dark matter of safety! Yet physics can confidently claim that dark matter and dark energy must exist, because the overall theories or models demand it. In safety science, assuming that it actually is a science (Hollnagel, 2014a), there is no comparable justification.

Skirting the issue (common and special causes)

The best way to solve the problems associated with the concept of a root cause is to avoid the issue altogether. This solution was chosen by Walter Shewhart, who had serious doubts about the existence of a root cause (as evidenced by his statement quoted later).

Common and special causes are the two distinct origins of variation in a process, as defined in the statistical thinking and methods of Walter A. Shewhart and W. Edwards Deming. Briefly, common (or non-assignable) causes, also called natural patterns, explained the usual, historical, quantifiable variation in a system (that we now tend to call performance variability), while "special (or assignable) causes" explained the unusual, not-previously-observed, non-quantifiable variation. But this does not completely avoid the root cause issue, since assignable causes can be attributed to something that happens or has happened in principle ad infinitum despite Shewhart's objections and therefore in principle can be eliminated or avoided, which is actually not too different from the Heinrich dogma. Proposing non-assignable causes is thus tantamount to admitting ignorance.

The logic of Safety-II and visio centum

Safety-II is simply defined as a condition where as much as possible goes well.

Safety-II and visio centum of course also include a set of assumptions primarily about why work usually goes well rather than about why it sometimes fails. The main difference between Safety-I and Safety-II is that accidents from the latter perspective are not seen as due to harmful factors and singular causes or conditions but that both acceptable and unacceptable outcomes should be explained in the same way. This obviously has important consequences for the practices of management and learning. The ethos of Safety-II is that there are desirable ways of behaving that should be encouraged and made easier, in contrast to Safety-I which holds that there are undesirable ways of performing that should be suppressed, constrained, or eliminated.

The logic of visio centum

The essence of visio centum is this: If everything goes well, then there will be no accidents.

1.   Therefore, try to make sure that as much as possible goes well by finding out how work goes well and learn from that, rather than look for the causes of failures!!!

Table I.2 How the world works according to visio centum and Safety-II

Accidents happen because	*"Nothing" happens and work goes well because*
People find ways to overcome design flaws and hindrances	People find ways to overcome design flaws and hindrances
People adjust what they do to match demands and conditions	People adjust what they do to match demands and conditions
People interpret and apply procedures to fit the situation.	People interpret and apply procedures to fit the situation.
People detect and intervene when it looks like something may go wrong	People detect and intervene when it looks like something may go wrong

2. To do so we need to understand how and why work goes well and pay attention to what happens when "nothing" happens and safety is present!
3. To manage safely we should try to make sure that as much as possible goes well by finding out how work goes well and learn from that, rather than from the causes of failures!!! Visio centum therefore requires a different focus of and approach to investigation. It becomes necessary to do so understand how and why things go well, or what happens when "nothing" happens ("nothings" are the non-events proposed by Weick and also what people say if no accidents have been reported for a period).

Managing safely

Describing what does not happen:

Investigating differently

It may initially seem difficult but it is by no means impossible to describe and investigate what did not happen. The feasibility of looking at something that did not happen or has not yet happened is demonstrated by the concept of pre-accident investigations (Conklin, 2012). It is also precisely what we do whenever we plan work – for ourselves or for others – and try to imagine a work situation in detail, such as deciding on the number of qualified people needed, as well as other types of resources and facilities, etc. Investigating differently also means that investigations and learning no longer are products of reporting, unless reporting is expanded to include everyday operations (Hollnagel et al., 2021).

The human craving for certainty

All human attempts to make sense of what happens around them is driven by a craving for certainty and a nearly universal need to avoid any feelings of insecurity. Every now and then this leads to a distortion or misinterpretation of facts, including a selection of which pieces of information are considered as facts in the first place. In psychology it is known as a confirmation bias or a tendency to seek out and prefer information that supports our preexisting beliefs, and sometimes also ignoring any information that contradicts those beliefs. The need for certainty has been described by philosophers and great writers. Great writers are usually also tolerably good philosophers, but the opposite relation is rarely the case; a few examples are provided in the following

Leo Tolstoy (1828–1910)

In his great novel, *War and Peace*, Leo Tolstoy discussed how the inhabitants of Moscow, the Muscovites, tried to make sense of the conflict when Napoleon was marching towards Moscow. Tolstoy remarked that

Man's mind cannot grasp the causes of events in their completeness, but the desire to find those causes is implanted in man's soul. And without considering the multiplicity and complexity of the conditions any one of which taken separately may seem to be the cause, he snatches at the first approximation to a cause that seems to him intelligible and says: "This is the cause!"

(Tolstoy, 2007, Book 13, Chapter 1)

Erich Fromm (1900–1980)

The famous American psychoanalyst said the same thing but in far fewer words: "The quest for certainty blocks the search for meaning."

William James (1842–1920)

William James was an eminent American philosopher and psychologist who believed that any number of impressions, from any number of sensory sources, falling simultaneously on a mind which has not yet experienced them separately, will fuse into a single undivided object for that mind.

The law is that all things fuse that can fuse, and nothing separates except what must.

(James, 1890, p. 488)

Here James points out that unless we are able to distinguish the details, we experience phenomena as a single entity (object or cause).

Walter Shewhart (1891–1967)

Walter Andrew Shewhart was an American physicist, engineer, and statistician, best known as the father of statistical quality control and of the PDCA cycle (described in Part IV).

As human beings, we want a cause for everything but nothing is more elusive than this thing we call a cause. Every cause has its cause and so on *ad infinitum*. We never get quite to the *infinitum*.

(Shewhart, 1931, p. 131)

Friedrich Wilhelm Nietzsche (1844–1900)

The great German philosopher, of course, also had an interesting contribution to make.

To trace something unknown back to something known is alleviating, soothing, gratifying and gives moreover a feeling of power. Danger, disquiet, anxiety attend the unknown – the first instinct is to eliminate these distressing states. First principle: some explanation is better than none. . . . The cause creating drive is thus conditioned and excited by the feeling of fear.

Nietzsche (1997, org. 1887, Chapter 5)

Thomas Hobbes (1588–1679)

The famous British 17th-century philosopher Thomas Hobbes wrote in the *Leviathan* that

Ignorance of remote causes, disposeth men to attribute all events, to the causes immediate, and Instrumentall: For these are all the causes they perceive.

Thomas Hobbes (1651, Chapter XI)

Managing safely

The essential problem with managing safety is that it is never safety as such that is managed, but rather the essential business of a company, an airline, for instance or a health care centre – the reason that safety is singled out as a separate issue is entirely a historical artefact (Hollnagel, 2020).

Even if you cannot directly manage performance by continuous adjustments and corrections as in managing a machine, such as the adaptive cruise control of a modern car, you can manage the potentials for performance, provided you have an articulated model of how the performance potentials actually shape or determine performance. Such a model must be developed as part of this effort and must be far more detailed than, for example, a safety culture model!

It makes sense because according to Mach's principle people do essentially the same thing in both cases. It is only the outcome that differs. Whenever someone begins to do something, he or she expects that the outcome will be acceptable as intended. Otherwise he or she would obviously do it differently. The expectation is based on a recognition of the situation and on previous experience (essentially a case of recognition primed decision making (Klein, 1998)). There are also far more cases when work goes well, and it is therefore much easier to study – there is no need to wait for an accident in order to learn something! "Erkenntnis und Irrtum fließen aus denselben psychischen Quellen; nur der Erfolg vermag beide zu scheiden." In English: "Knowledge and error flow from the same mental sources, only success can tell one from the other" (Mach, 1908).

The reification fallacy

Reification is the mistake of treating something abstract as if it was concrete, as if it was something that existed independently of the description. Safety is based on a collaborative consensus rather than on observations of something physical. It is, as James Reason (2000) pointed out, defined by its absence rather than by its presence. Safety is the result of a social convention and is therefore a social construct. People within the same community therefore tacitly agree on what it means. The term is unfortunately so common and used so often that no one pays attention to it any longer. Referring to safety as if it was something is therefore an example of reification. Using safety as a noun (mis)leads us to talk about safety management, thereby implying that safety by itself is something that could be managed. But safety is not a something. Being safe instead describes a characteristic of a state or of how something happens, meaning either that as little as possible goes wrong (as in Safety-I and vision zero) or that as much as possible goes well (as in Safety-II and visio centum). Rather than managing the safety of operations the concern should be how to manage operations safely, where safely is used as an *adverb* rather than as a *noun*.

What is safety?

It is not uncommon to see this question and it is easy using Google Scholar, for instance, to find publications that aim to provide an answer. No matter what the answer is, they all fail to recognise that asking the question represents the reification fallacy and therefore is a mistake, because it implies that safety *is* something, The proper question to ask is "what does it mean to be safe?" "Being safe" is a state or condition, defined either as vision zero or as visio centum. And the interesting question is less what it is and more how it can be achieved. An answer to that has practical value unlike an answer to "What is safety?"

Legacy epilogue

Part I has given a characterisation of how people and organisations in practice consider safety by looking at the ways in which they conventionally try to manage safety. The essence of this perspective is captured by the safety mantra and by the Heinrich dogma. It can be considered a kind of safety legacy although it is more like a mental straitjacket that constrains how we think about safety and therefore also how we deal with the issue in practice. The argument put forward by this text is that we traditionally investigate what has happened – in most cases because it did not go as intended and expected – and that this attitude has become so widely accepted that it is never doubted or questioned.

With this entrenched focus on adverse events it is no surprise that the primary aim is to reduce their number. The main lessons learned are therefore about what should *not* be done and about what one should try to *prevent or avoid* in the future. Another possibility, which represents a contemporary systemic (Safety-II) way of thinking, is to focus on what was intended and expected, what *should* have happened but for some reason did *not* actually happen. In this way the main lessons learned are about what should be *done* and what one therefore should try to *promote* and *facilitate* in the future, individually as well as organisationally; this approach to investigation and learning is not only more constructive, but also more motivating for everyone involved, hence easier to implement and sustain.

II   The complexity conundrum

Complexity prologue

There can from time to time be a gap between our ability to understand systems and our ability to manage them. We habitually account for this gap by calling the systems complex, in the vain hope that someday there will be a magic formula that miraculously overcomes the complexity issue. But we might instead consider what the problem really is and try to solve it. In this context our Achilles heel is the temptation to rely on the linear cause-effect thinking that characterises the safety legacy described in Part I even though it is spectacularly unable to account for the non-linear phenomena (unanticipated disproportionate outcomes) that haunt socio-technical systems of today. A better but harder solution is surely to find an alternative to linear thinking.

Part II takes a closer look at complexity. The term has been used with increasing frequency in an almost defeatist manner in discussions about contemporary society, not least in relation to major unwanted events, accidents, and other disasters (including administrative blunders). The concept has, for instance, given rise to the often-used label Complex Adaptive System (CAS), where complex means that the system consists of many interconnected parts and adaptive means that it can change its performance according to conditions which make it unmanageable in principle as well as in practice. A CAS is therefore by definition impossible to control or manage. But CAS is merely a convenient label (calling a complex system complex is not very helpful and does not actually explain anything).

The rise of complexity

The use of the term *complex* in relation to safety probably began in 1984 when Charles Perrow proposed the existence of normal accidents and argued that accidents by that time – 40 years ago – should be seen as normal or even inevitable rather than as exceptional occurrences because society and the systems and organisations we have made ourselves dependent on to sustain our comfortable way of life had become incomprehensible. Perrow explained complexity analytically using the concepts of *couplings* (either tight or loose) and *interactions* (either simple or complex) and illustrated these by the interaction/coupling chart in

DOI: 10.4324/9781032664729-2

Figure II.1. It is an interesting exercise to consider what it would look like today, 30 years later, based on what has happened since 1984 – the financial system and health care would probably feature prominently (but I leave this exercise to the reader). Tight couplings mean that parts and functions are highly dependent on one another, and complex interactions are plainly complex, with no further definition provided. The effects of failures, disruptions, and performance variability can therefore quickly propagate through the system to produce unexpected and non-linear (disproportionate) effects. Tracing the origin to "normal accidents" raises the question of whether complexity is a trait of the description or a trait of that which is being described as discussed by Pringle (1951). In other words, is complexity epistemic, having to do with what we can know about something and how it can be described? Is the origin from *epistēmē*, the Greek word for knowledge? Or is it ontological, having to do with the nature of being or the nature of what is described?

Systems with tight couplings and complex interactions (in the second quadrant of Figure II.1) are therefore to all extents and purposes difficult to describe and understand. (One possible explanation for the increasing complexity of our socio-technical systems is the *circulus vitiosus* shown in Figure II.2). Complexity is, however, an emergent outcome rather than a result, where the impetus comes from the never-ending technological innovations, that are eagerly used to improve our undertakings (ostensibly to make them faster, better, and cheaper

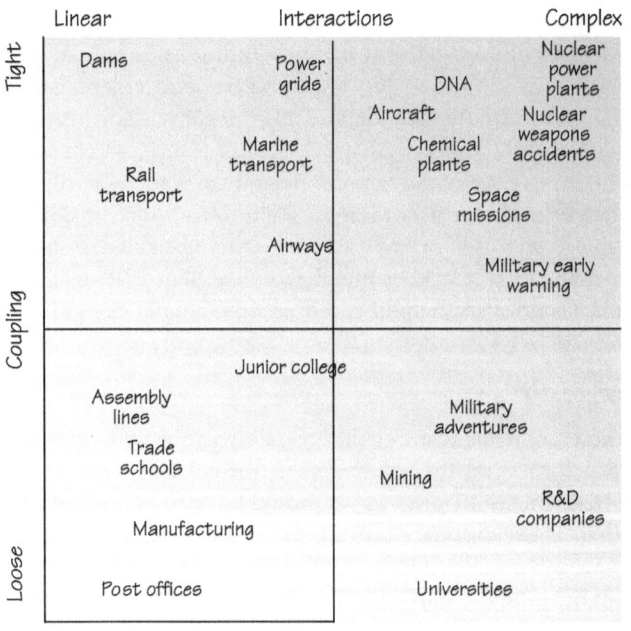

Figure II.1 Coupling-interaction chart (from Perrow, 1984, p. 327)

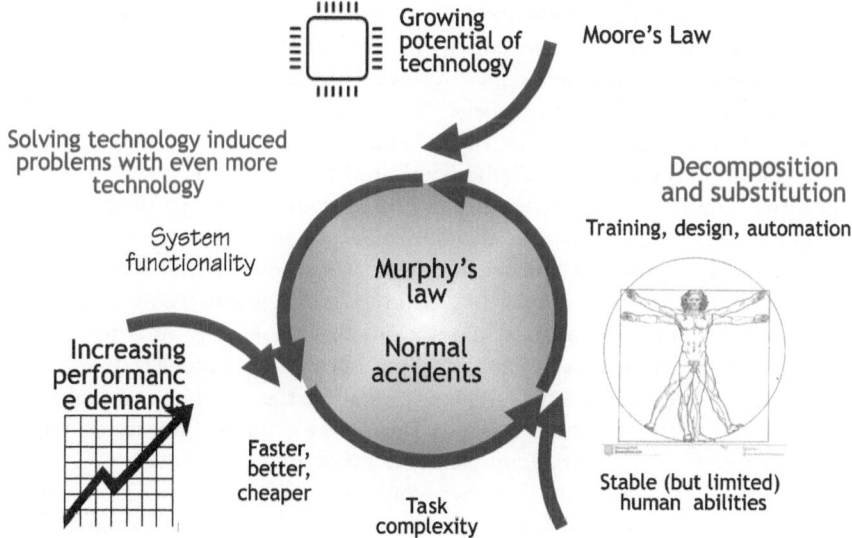

Figure II.2 The forces driving the increasing complexity of socio-technical systems

(Paxton, 2007)) but ironically also to compensate for the problems created by the precipitous use of technology to solve other problems in the past.

Part II will therefore take a closer look at complexity. The term has been used with increasing frequency in an almost defeatist manner in discussions about contemporary society, not least in relation to major unwanted events, accidents, and other disasters (including administrative blunders). The concept has, for instance, given rise to the often-used label *Complex Adaptive System* (CAS) where *complex* means that the system consists of many interconnected parts and *adaptive* means that it can change how it performs to match the conditions, which makes it unpredictable in principle as well as in practice. A CAS is therefore by definition impossible to control or manage. The use of the term *complex* in safety can be traced to the year 1984 when Charles Perrow introduced the concept of normal accidents and argued that accidents by then should be seen as normal or even inevitable rather than as exceptional occurrences because society and the systems and organisations we have made ourselves dependent on to sustain our comfortable way of life slowly had become incomprehensible. Perrow explained complexity using the concepts of *couplings* (either tight or loose) and *interactions* (either simple or complex) and illustrated them by the interaction/ coupling chart in Figure II.1. Tight coupling means that parts and functions are highly dependent on one another. The effects of failures, disruptions, and performance variability can therefore quickly propagate through the system and lead to effects that are both unexpected and non-linear (disproportionate). The systems are therefore to all extents and purposes complex to behold and therefore also difficult to manage. And complex interactions are evidently complex, with no further definition provided. Systems with tight couplings and complex interactions are therefore to all extents and purposes difficult to describe and understand and therefore also difficult to manage. (The British philosopher George Henry Lewes had, however, already in 1877 during the debates following Darwin's theory of evolution contrasted "emergent" and "resultant" effects and noted that the former are neither additive nor predictable from knowledge of the system's components nor decomposable into such components.) According to Perrow these systems are prone to "normal accidents." One possible explanation for this development is the *circulus vitiosus* shown in Figure II.2, where the impetus comes from never-ending technological innovations that are eagerly used to improve work (ostensibly making it faster, better, and cheaper (Paxton, 2007)) and incidentally also to compensate for problems created by the precipitous introduction of technological fixes to other problems. Tracing the origin to "normal accidents" raises the question of whether complexity is a trait of the description or a trait of that which is being described. In other words, is complexity epistemic, having to do with what we can know about something and how it can be described? Or is it ontological, having to do with the nature of what is being described (Pringle, 1951)? Or simply, is complexity real or is it a postulate?

The modern use of complexity

The current infatuation with complexity began in the 1990s. Whenever the term was used prior to that, it usually simply meant complex, without any metaphysical overtones (e.g., Beer, 1959). In an introduction to a special issue about the subject, Urry (2005) remarks that: "More generally, the term 'complexity' is 'present' and doing metaphorical, theoretical and empirical work within many social and intellectual discourses and practices besides 'science'" (p. 2). These other "intellectual discourses and practices" span from alternative healing to town planning, with many interesting topics in between. Even if we limit ourselves to the scientific uses of the term, it remains ill-defined. "Complexity investigates emergent, dynamic and self organising systems that interact in ways that heavily influence the probabilities of later events. Systems are irreducible to elementary laws or simple processes" (Urry, 2005, p. 33). The assertion that systems are irreducible to elementary laws or simple processes is practically an expression of faith, since it is impossible to either prove or falsify. While it makes sense to shun reductionism as a universal principle for scientific explanation, the fact that we at present are unable to "reduce a system to simple processes" does not mean it will always be the case. To assume that would go against a philosophical principle that is more than 700 years old. The term or label '*complexity*' is increasingly common in papers and books that relate to human performance in socio-technical systems and indeed to the systems themselves. The term does not appear in Buckley's collection of papers *Modern Systems Research for the Behavioral Scientist* Buckley (1968).

The preference for monolithic explanations

Even though we readily admit that the problems we face are increasingly complex, we do our best to avoid complexity when it comes to thinking about, describing, and understanding them. For some reason we tenaciously cling to the belief or hope that even complex phenomena and systems can be described in simple ways – by linear causality and pairwise cause-effect relations – and with simple terms, preferably single and uncomplicated concepts. But we must never forget that while simple problems possibly may have correspondingly simple solutions, complex problems practically always require correspondingly complex solutions. Disguising complex problems as simple problems by offering apparently simple solutions does not make the problems any simpler; a more likely consequence is that the solutions will not be effective. But although effects or outcomes usually can be described by simple terms and concepts, such as the HAZOP guide words in Table II.1, it cannot be taken for granted that the same goes for their underlying causes. In terms of evolution, the human mind and comprehension lag considerably behind the socio-technical systems that mind and body have to cope with. Humans prefer explanations that rely on a single concept or factor, here called monolithic causes. As social constructs, monolithic causes are efficient (easily found and accepted) but lack thoroughness and precision. A monolith is a geological feature consisting of a single massive stone or rock. A well-known example of a monolith is the black slab in Stanley Kubrick's film *2001: A Space Odyssey*.

Complexity is a monolithic concept, just as safety is.

Monolithic explanations are convenient because they point to a single and therefore uncomplicated cause that sometimes too easily can be associated with similar monolithic or simple antidotes or solutions (the much sought-after "silver bullets"), even though they are likely to introduce unanticipated consequences. Relying on monolithic explanations and resorting to monolithic solutions is an intellectual shortcut and a trade-off between efficiency and thoroughness that is convenient but risky. The risk is that problems are defined in terms of whether

Table II.1 Typical HAZOP guide words

HAZOP guide word	Meaning
No (not or none)	None of the intended outcomes were achieved
More (more of or higher)	Quantitative increase of the outcome
Less (less or lower)	Quantitative decrease of the outcome
Part of	Only some of the intended outcomes were achieved
Reverse	The outcome was the opposite of what was intended.
Other than (other)	Complete substitution (the outcome is completely different from what was intended)

they correspond to familiar monolithic solutions, often involving some form of new technology; see Figure II.1. Yet technology is in most cases a solution looking for a problem and is usually adopted because it seems advantageous to do so rather than because there is a well-defined need.

Common and popular examples of monolithic causes and their antidotes are listed in Table II.2:

Table II.2 Monolithic causes and their antidotes

Monolithic cause (as a social construct)	Monolithic solution or "silver bullet"
Technology	Redesign, construction, maintenance
Human error	Prevention, elimination, compliance, standards
Counterfactual condition (lack of) X (e.g., trust, safety culture)	Provide or improve X (whatever was missing or deficient)
Deviations from norms	Standardisation, compliance

Whether we like it or not, most of us are what the father of Cybernetics, Norbert Wiener, called *gadget worshipers*, which he defined as people who:

> regard(ed) with impatience the limitations of mankind, and in particular the limitation consisting in man's undependability and unpredictability.
>
> (Wiener, 1964)

Today Wieners' concept of gadget worship has practically become a design philosophy of its own, called *solutionism*, defined as "An intellectual pathology that recognizes problems as problems based on just one criterion: whether they are 'solvable' with a nice and clean technological solution at our disposal" (Morozow, 2013a, 2013b).

The preference for single and simple explanations is ubiquitous in how humans strive to understand what goes on around us. It can be found in all fields of activity, politics, ethics, law, biology, history, finance, science – and of course in industrial safety. In the latter case it is convincingly illustrated by the categorisation of safety thinking as having three ages called the age of technology, the age of human factors, and the age of safety management, respectively; see Figure II.3 (Hale & Hovden, 1998). This categorisation highlights the fact that for each age a single explanation or cause served as a solution to a variety of problems. The explanations were in chronological order: technical failures or malfunctions, "human error," or the human factor and organisational culture, and the corresponding solutions were "repair and improve," "blame, followed by train, design, and automate," and "safety culture," respectively. Such simple solutions are obviously attractive since they all make it easier to explain what has happened and to communicate it to others. While their practical value in most cases is limited, their emotional value, their ability to set the

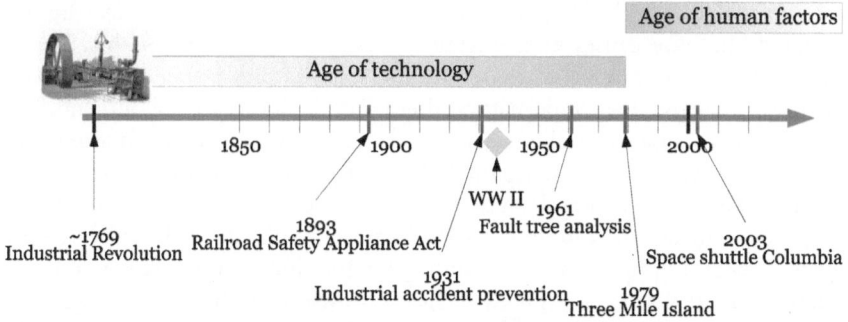

Figure II.3 The three ages of safety based on Hale and Hovden (1998)

mind at ease, is priceless. Explanations of this nature can be called monolithic because they rely on a single concept and in the sense that they constitute a single unit. Relying on these in the choice of responses is also likely to introduce unanticipated consequences.

The three ages of safety

Monolithic causes can be seen as representing a social convention and therefore as being social constructs. They can also be seen as a form of an efficiency-thoroughness trade-off (Hollnagel, 2009). Monolithic explanations are efficient to use, quick to apply, and require little effort – cognitive, mental, or otherwise. Monolithic explanations or causes are not constant and universal but must follow the changes in the age of safety, changes to the dominating way of thinking (Hale & Hovden, 1998). The three ages introduced by Hale and Hovden (1998) are shown in Figure II.3. But the causes lack in thoroughness and in precision. This will sooner or later show itself as an inability actually to improve the situations where they were used.

It is not really surprising that monolithic thinking is so pervasive, since we can see the influence in the very language that is used to describe our mental activities. We do talk about a (single) line of thought or a (single) line of reasoning. Indeed, the classical ideal of reasoning as logical thinking (syllogisms) is itself strictly linear.

The collapse of the Rialto bridge (1444)

People did of course not begin to think about safety only after the industrial revolution had started. People must have been concerned even before that whenever something went wrong leading to losses, injury, and perhaps death, despite a lack of documentation and although the general understanding of how the world worked was less developed. So we may propose an age – or era – before the age of technology that somewhat arrogantly will be called the state of beliefs, because causes were attributed based on beliefs rather than knowledge, typically either to Nature or to some supernatural power or God (but it might not have seemed supernatural then). The belief that developments were determined by supernatural powers can be seen from the practice of looking for signs, for instance by observing the behaviour of birds and sometimes even to sacrifice them to take signs from the state of their entrails. (Why the gods would use this as a means of communication was never explained.) Shakespeare's play *Julius Caesar* is full of examples of looking for and interpreting dreams and signs. Caesar himself was even reputed to have thrown a die before he crossed the Rubicon River in 49 BCE on his way back to Rome and afterwards presumably uttered the famous words "*alea iacta est*" (the die is cast). Indeed, it was during that stage generally assumed the higher powers determined the outcomes even when gambling with dice. The concept of randomness did not exist; even the concept of a fair die did not exist (neither Nathan Detroit nor Sky Masterson would have appreciated that). Anything that happened was a result of God or the gods making it happen. And without a concept of randomness, the only possible explanation for whatever happened was, of course, strict causality. This belief remained until the Franciscan friar and mathematician Luca Pacioli (1445–1517) in 1494 published the first ever text on probability (*Summa de arithmetica, geometria, proportioni e proportionalita*). Pacioli is also known as the father of accounting and bookkeeping. Based on Pacioli's work, the Renaissance gambler, mathematician, physician, and astrologer Gerolamo Cardano later in 1564 wrote the first ever book on probability and gambling *Liber de ludo aleae*. (English title: *The Book on Games of Chance*). Cardano lived from 1501 to 1576. The book was, however, not published until 1663, so Blaise Pascal and Pierre de Fermat would not have known about it when they produced their own work in 1654 almost a century later.

In the same way there must be an age – or stage – following the third age of safety management, which we can call the age of complex systems, the one that we live in now. During the first stage of beliefs, one spectacular pre-industrial accident is the collapse of the Rialto bridge in Venice in 1444 when it became filled with spectators at the wedding of the Marquess of Ferrara. (There must have been many others, but documentation is scarce.) There must have been some head scratching and speculations about why this happened, as well as an attempt to blame someone, but probably no accident investigation as such. If it was generally agreed that the accident was due to the will of higher powers, there would, of course, be no reason to investigate it. It would most likely have been ascribed to some kind of higher powers, if not specifically to the displeasure of God – but surely not at a wedding between nobility! (With the knowledge we have today, a more likely explanation is that the bridge became overloaded by the many spectators.)

Another example is that the region of Tuscany suffered from a serious plague epidemic between from 1631 to 1633. The epidemic ended in September 1633. Grand Duke Ferdinand attributed this to the May procession of the Miraculous Madonna of Imprunenta. Compare that to the cholera epidemic in London in 1854 where Dr John Snow through an early use of statistics traced the source to a pump in Broad Street. On September 7, Dr Snow persuaded the Board of Guardians of St. James's parish to remove the handle of the pump. This happened the following day and effectively prevented further spreading of the cholera, something the procession of a saint would never have achieved.

The capsizing of the Vasa *(1628)*

Another spectacular accident but much better researched and documented is the capsizing of the warship *Vasa* in 1628 (Flin, 2006)

On August 10, 1628, crowds of people stood on the quays of Stockholm harbour as the grandiose royal warship *Vasa*, after three years of construction, set out on her maiden voyage to join the Swedish navy.

The *Vasa* was no ordinary warship. It was built during the Thirty Years' War in Europe, and was intended to be used against Poland. King Gustavus II Adolphus Vasa wanted her to be the mightiest in the world. Some say that he ordered a second gun deck built after he heard that the Danes were building a ship with two gun decks. He wanted the ship that carried his family name to be inferior to none.

Her departure was supposed to be a showy display of his royal power and glory. She was armed with 64 guns and adorned with more than 700 sculptures and ornaments. Her price was more than 5 percent of Sweden's gross national product at the time. This powerful war machine and floating art exhibition was probably the most glorious ship built anywhere at that time. No wonder people were cheering her on with pride as she passed the quays of Stockholm!

However, the *Vasa* had sailed less than one nautical mile (little more than one kilometre) when a strong gust of wind made her heel over. Water gushed in through the open gun ports, and down she went. This must surely have been the shortest maiden voyage in naval history!

The spectators were stunned. The glory of the Swedish Navy was brought down, not in battle (by the dreaded Danes) or by a violent storm on the high seas but by a simple gust of wind in Stockholm's inner harbour. The death of about 50 people on board caused further consternation. Instead of being an object of national pride, the *Vasa* became synonymous with disappointment and disgrace.

A court was summoned to find the someone responsible for the humiliating catastrophe. But no one was charged, likely because the testimony implicated both the king and the second highest commander in the Swedish navy, Vice Admiral Klas Fleming. The Dutch ship builder was unfortunately no longer alive. His widow still resided in Stockholm, but she was never charged

The king's demands had made the builders experiment with designs unfamiliar to them. Thus, the *Vasa* became badly proportioned. Sometime before the capsizing, Admiral Fleming had arranged a stability test. Thirty men ran abreast from one side of the ship to the other. After three runs the admiral realised that if they continued, the ship would capsize right then. So he halted the test but strangely did not cancel the maiden voyage. With such important personalities as the king and the admiral implicated, the charges were dropped.

In 1664–1665, an ex-officer of the Swedish army recovered most of the *Vasa's* guns by means of a simple diving bell. The *Vasa* was then gradually forgotten as she sank deeper and deeper into the mire 100 ft (30 m) below the surface.

In August 1956, an amateur archaeologist, Anders Franzén, used a core sampler to bring up a piece of oak from the bottom. For years he had been examining old documents and searching the seabed looking for the *Vasa*. Now he had found her. Through a delicate salvage operation, shown live on TV, the *Vasa* was lifted out of the mud and carefully carried underwater in one piece to a waiting dock.

Five years later on April 24, 1961, the quays in Stockholm were again filled with cheering spectators. After 333 years at the bottom of the sea, the *Vasa* made her comeback – this time as a tourist attraction and a treasure for marine archaeologists. More than 25,000 artifacts revealed fascinating details about this 17th-century warship and also gave unique insight into contemporary shipbuilding and sculptured art.

Why were the *Vasa* and her artifacts so well preserved? Some factors were that she was new when she sank, the mud had a preserving effect, and the wood-destroying sea worm does not thrive in water with low salt content. Stockholm lies at the eastern end of the Mälaren lake system; the water in the harbour is therefore not very than salty. The salinity of the Baltic Sea is low anyway.

The *Vasa* had some 120 tons of ballast (rocks were commonly used at the time). Experts much later calculated that she needed more than twice that amount to make her stable, but she did not have the space. Also, such added weight would have brought the lower gun ports below the water line. That would have made her maiden voyage impossible! Her appearance was glorious, but her poor balance made her destined for disaster.

Now, as the oldest preserved, complete, and fully identified ship in the world, she is safe inside her own museum. The are 900,000 visitors a year get a glimpse of 17th-century royal ostentation, frozen in time by that catastrophe in 1628. It is a reminder of the folly of those in authority who, through ego and carelessness, chose to ignore sound shipbuilding practices.

The *Vasa* is well worth a visit the next time you are in Stockholm.

Complexity as a social construct

Classifying something as a social construct (Searle, 1995) simply means that it is based on a collaborative consensus rather than on observations of physical reality and that it represents a set of ideas shared by a number of people: safety, risk, and complexity are all good examples of social constructs, since they have no obvious physical reality (accidents as the absence of safety may have), unlike the Eiffel tower, which is there for all to see, touch, and climb. But a social construct is different from the psychological phenomenon called intersubjective verifiability (which is the characteristic of a concept or a phenomenon to be readily and accurately communicated between different individuals, who in that way can agree that they are indeed referring to, thinking of, or talking about the same phenomenon). Intersubjective verifiability is, however, different from naïve realism). In the case of complexity the consensus primarily rests on the faith in reverse causality, namely that there must be a cause for any observed outcome or consequence, plus the need for certainty that both motivates and constrains human reasoning. Complexity as a cause represents a satisfactory explanation for why an accident happened, and if enough people agree it becomes the undisputed and socially accepted cause. Social constructionism is in many ways similar some ways to the concept of a common basis of understanding used by phenomenological psychology. This term recognises the fact that when we try to understand something we repeatedly ask why, not unlike the "five whys" in the Toyota Production System. Yet the asking of why does not stop after five repetitions and neither does it, as in children, go on forever, but it stops when it reaches the common basis of understanding, something on which we (tacitly assume) we all agree where the answer is "obvious" to everyone who belongs to the same culture (whether it is the organisational culture or the same safety culture).

Complexity issues

Can complexity be idiosyncratic, and can something be complex to only one person but not to others? This is not too different from asking whether complexity is epistemic (a feature of the description of a phenomenon) or ontological (a feature of the phenomenon as such with some physical reality). Since complexity is social construct it can obviously not be idiosyncratic. For something to qualify as a social construct several persons can easily agree that they have the same experience.

Complexity characterises the behaviour of a system or model whose components interact in multiple ways and follow local rules, leading to non-linearity, randomness, collective dynamics, hierarchy, and emergence. The adverb *complex* was used 89 times in the report of the Columbia accident investigation board (CAIB; Gehman, 2003), for instance in the following conclusion: "complex systems almost always fail in complex ways." This makes perfect sense provided you already know what complex means. If we replace complex with "difficult to describe and understand," the conclusion becomes "systems that are difficult to describe and understand almost always fail in ways that are difficult to describe and understand," which is simpler and perhaps more honest and meaningful. For some reason, the noun *complexity* was used just 22 times.

Etymology

Like so many other words we use daily, *complex* has its roots in Latin *complexus*, past participle of *complecti, complectere*, to embrace. The meaning of complex as "involved, intricate, complicated, not easily analysed" is found as early as 1715. The last part of the definition "not easily analysed" agrees with the argument that complexity is epistemic rather than ontological. This means that complexity is a quality or characteristic of the description of a system rather than of the system as such. This is easy to illustrate. Most of us would presumably if put in the pilot's seat in a modern fly-by-wire airplane (whether from Boeing or Airbus, or even a DC-3) agree that it was complex, but an experienced pilot hopefully does not feel the same way. Similarly, if a person who had only ever driven a Ford Model T unexplainably was transported through time and put in a modern car, say the one we drive daily, he would also find it very complex. As would we if we were asked to drive a Ford Model T, not because it is complex as such but because we would not be familiar with it.

Coping with complexity

The concept of coping with complexity was introduced by Rasmussen & Lind, 1981) and later became one of the three main themes of Cognitive Systems Engineering (Hollnagel Woods, 1983). The two other themes were joint cognitive systems and the use of tools (Hollnagel & Woods, 2005; Woods & Hollnagel, 2006). But what did complexity mean in the 1981 publication? It referred to complex situations during plant disturbances, hence that the ways in which a disturbance could develop and escalate were non-trivial and difficult to understand, predict, and manage. A good example of how we all cope with complexity is provided by the characteristic responses of people who encounter a condition of Information Input Overload (described later).

Complexity as a monolithic cause

Even though the terms *complex* and *complexity* are used with increasing frequency, it is not because complexity represents something attractive. No one in their right mind will intentionally build or create a complex product or service. Complexity is something we try to avoid, especially when we have to explain something. That we make the world complex anyway is perhaps best explained as cognitive lethargy. So even though many of the problems we encounter and try to describe and understand are complex, we strive to make the descriptions themselves simple as if that by magic also would make the problems simple, except when we give in and call them complex, although using complexity as an explanation actually is a gross simplification, because it is a monolithic explanation; we usually describe why something (unwanted) has happened, either by accounting for its presence as a hypothetical cause (examples are "human error" and complexity) or by accounting for their absence as a critical or counterfactual condition (examples are most non-technical skills such as situation awareness or leadership).

Complexity is currently the latest in the sequence of monolithic explanations. But the very idea of a sequence ironically assumes linearity, even though complexity itself represents non-linearity.

Complexity is used as a name for that which is hard to understand, which ironically includes the concept of complexity itself. But there are other, more operational terms available, although whatever complexity refers to may be just as hard to understand, if it is called something else. One alternative set of terms is *tractable* and *intractable*.

There are two pairs of terms that may serve as alternatives to complexity (tractable/intractable, and trivial/non-trivial). Something is tractable if it is easy to follow, control, and influence.

Complexity is without a doubt epistemic rather than ontological. Because if complexity was ontological, then it would in principle be possible to give a simple description of it.

Alternative terms to complexity (1): tractable and intractable

A simple definition is that a system is intractable if it changes during the time it takes to comprehend it. We try to design work situations that are tractable, indeed, situations that can be grasped intuitively and which hopefully do not change at all. This is also the reasoning behind quality management and quality assurance. But every now and then our noble efforts fail, and unpredictability creeps in.

Due to technological developments, workplaces at one time reached a point when systems no longer were tractable. We may call this a singularity. (In mathematics, a singularity is in general a point at which a given mathematical object is not defined.) Systems ceased being tractable when tight couplings became necessary for them to function as required. In industry this probably happened when microchips became an integral part of machines of all kinds (although there is no unique event that marks that, a reasonable guess is around 1975, when Gordon Moore formulated what later became known as Moore's law).

Table II.3 Characteristics of tractable and intractable systems

Characteristics of tractable systems	Characteristics of intractable systems
Description are simple with few details	Description are elaborate with many details
Principles of functioning are known	Principles of functioning are incompletely known
The system is stable and does not change while being described	The system is dynamic and may change before a description has been completed.

Alternative terms to complexity (2): trivial and non-trivial

Human factors can itself be seen as a reply to a state of intractability; for example due to the rapid introduction of new technology it became impossible for people to adapt themselves to the systems, because they were non-trivial. This means that the systems became unpredictable and therefore also uncontrollable.

Many or most of the methods that we use have been developed for tractable and trivial systems and are therefore not adequate for the intractable and non-trivial system we work in today.

In order to cope with the complexity around us it is necessary for people to adjust what they do to the conditions, in other words rely on the performance variability that Taylorism and human factors engineering tried so hard to wipe out. Human behaviour is always varying for a number of reasons (individual reasons are psychological and physiological, but there may also be other reasons that have to do with the workplace and are more social (like norms) and shared attitudes and values). But over and above that it becomes necessary to adjust performance approximates when the world and work environments no longer are sufficiently predictable. This is demonstrated in decision making where descriptive theories of decision making such as satisficing, (March & Simon, 1958; Simon, 1947), muddling through (Lindblom, 1959), and naturalistic decision making (Klein, 1993) gradually and irresistibly have replaced the normative theories of rational economic decision making, despite heroic attempts to maintain a resemblance of rationality (Tversky, 1972).

The terms *trivial* and *non-trivial* were proposed by Heinz von Förster who, together with Warren McCulloch, Norbert Wiener, John von Neumann, and others, laid the foundation for cybernetics. In particular von Förster developed a second-order cybernetics which focused on self-referential systems and the importance of *eigenbehaviors* or the "cybernetics of observing systems" for the explanation of complex phenomena, Self-referential systems are related to self-organising (autopoietic) systems which can turn into the dreaded Complex Adaptive Systems. Many years later Maruyama (1963) used the term *second-order cybernetics* to describe the *Deviation-Amplifying Mutual Causal Processes* that Jay Wright Forrester later used in his system dynamics world models. Von Förster's concept of a non-trivial system provides a good alternative to complexity. In a trivial system, current operations are not influenced by previous operations. A trivial system is analytically determinable and thus predictable. This is, however, not the case for non-trivial systems, where the problem of identification, for example deducing the structure of the system from how it performs, is unsolvable. Non-trivial systems are therefore complex, for example the quotation from Bernard Russel in Part I.

The problem is, in von Förster's words, related to the "cybernetics of observing systems." Von Förster's concept of a non-trivial system is a starting point to recognise the complexity of cognitive behaviour. Many of the problems we face in safety (and resilience) are due to non-trivial systems that are not analytically indeterminable. In such systems what happens next depends on what happened before, and they are therefore unpredictable, except to de Laplace's demon.

Potential sources of complexity

The law of unintended consequences

A possible explanation for the continuously growing complexity is Robert Merton's "law of unanticipated consequences." In 1936 Robert K. Merton, who was an American sociologist considered as the founding father of modern sociology, published a pioneering paper with the intriguing title "The unanticipated consequences of purposive social action." Here Merton put forward the following five reasons or explanations for why decisions often have unanticipated consequences:

Table II.4 The law of unanticipated consequences (from Merton, 1936)

Reason why a decision may have unanticipated consequences	*Explanation*
Ignorance	either as general inadequate knowledge about consequences of actions and decisions or specifically because the decision maker has an incorrect, stochastic (rather than functional) organisation model.
Error	plain and simple, as in false reasoning or reasoning with limited scope (due to negligence or bias).
Imperious immediacy of interest	Meaning that a dominant interest in primary/ immediate results leads to a neglect of possible side-effects
Basic values	As when the importance of basic values (criteria) means that long-term consequences are neglected
Self-defeating predictions	When one's own predictions become a new element of the situation, thus tending to change the initial course of developments

Although Merton did not mention it specifically, it seems plausible that such unanticipated consequences frequently occur in the context of choosing remedial actions after a major accident where the demands to speed and efficiency dominate any concerns for thoroughness and that this may be one of the main reasons for the ever-growing complexity, not just for safety management in socio-technical systems and in the managerial oversight but also for responses by regulators and legislators.

When introducing responses to events, as antidotes in the case of safety, efficiency (speed) is often rated higher than thoroughness, because of urgently felt needs (public or political). This may lead to responses that are premature or incomplete and therefore produce the unintended consequences that Merton described. And the occurrence of unanticipated consequences may make the results non-trivial and more difficult to comprehend and in this way contribute to both the epistemological and the ontological (real) complexity.

Complexity due to information input overload (IIO)

One way of explaining why something is complex is to suggest that there simply is too much information, either in an absolute sense, as suggested by Rouse and Rouse (1979) or in a relative sense, such as more than can be comprehended, digested, or processed in the situation. Even psychologists who do not accept the

Table II.5 Typical responses to Information Input Overload (IIO)

IIO Strategy	Justification
Omission – or a temporary non-processing of information; this response is used if it is important to complete the task without further disturbances.	It is important to complete the task without further disturbances.
Reduced precision – trading speed for time.	It is important to reduce or compress time but not miss essential information.
Queuing – delaying response during high load, if it is important not to miss any information (this is only efficient for temporary conditions and based on the hope that things will become less stressed later). And it is always done with an optimistic hope of catching up later.	Delaying response during high load on the optimistic assumption that it will be possible to catch up later (stacking input).
Filtering or cutting categories – neglecting to process certain categories if time/capacity restrictions are severe and it is deemed sufficient to notice only large variations.	If time/capacity restrictions are really severe and it is sufficient to note only large variations.
Parallelisation – if sufficient resources or capacity are available, information can be processed in parallel	Distributing or delegating processing if possible; calling in assistance.
Decentralisation delegation – if sufficient resources or capacity are available, information can be processed in parallel.	Trying to deal with incoming information as well as possible
Escape – abandon the task altogether. The ultimate solution which completely removes the overload but at the cost of not completing the job at hand.	Self-preservation

idea promoted by the likes of Lindsay and Norman (1972), Newell and Simon (1963, 1972), and Newell (1990) that humans are just information processing systems, although possibly complicated ones (leading to a plethora of simple box-and-arrow (information flow) models), will have to admit that there are limits to our attention and retention and to how much information we can cope with and how fast. At one point in time, the discussion centred around whether the limitations should be explained as epiphenomena of channel capacity or of processing capacity (Moray, 1967 and also Wickens, 1992) in line with the traditional academic approach to explain something new by invoking known concepts, models, and theories, to determine a scientifically plausible cause. A radically different approach was to notice what people actually did to cope with the problem which became known as Information Input Overload (Miller, 1960). These observations revealed that there are a limited number of ways in which people cope with an overload of information.

Monolithic explanations are the ultimate form of filtering: they collapse the number of categories to just one – "1" – it therefore eliminates any need for mental work and categorisation/discrimination. Unfortunately, it is also an efficiency-thoroughness trade-off of the worst kind, since it makes it impossible to notice any differences.

Complexity due to Information Input Underload (IIU)

It is quite ironic that some of the responses to an overload of information actually may reduce the input so much that the result is the opposite: a condition of too little information or Information Input Underload (IIU).

Information Input Underload may occur either if some information is missing (a *true* underload condition) or if it has been discarded for one reason or another, for instance in response to a preceding overload condition leading to a *relative* underload condition.

Not having sufficient information can, of course, also contribute to complexity – if there is not enough information to understand what goes on and to follow how an event or a process develops they will be difficult to manage and control. Although IIU has not been studied to the same extent as IIO, it is still possible to identify some of the typical strategies people employ to cope with the condition. They are:

> Extrapolation: Here existing evidence is "stretched" to fit a new situation; extrapolation is usually linear and is often based on fallacious causal reasoning.
>
> Frequency gambling: Here the frequency of occurrence of past items/events is used as a basis for recognition/selection.
>
> Similarity matching: Here the subjective similarity of past to present items/events is used as a basis for recognition/selection. Similarity matching and

frequency gambling were both described by (Reason, 1990a) but are in many ways similar to the representativeness and availability heuristics proposed by (Tversky & Kahneman, 1974).

Trial and error (random selection): Here interpretations and/or selections do not follow any systematic principle.

Laissez-faire: Here an independent strategy is given up in lieu of just doing what others do. The effect is similar to the escape response for IIO.

The problem of Information Input Underload has received even less attention than Information Input Overload. One possible reason is that overload conditions are more conspicuous and that overload also is a more striking condition than underload, as well as easier to create experimentally (one attempt was made by Marshall et al., 1981). From a qualitative point of view, Information Input Underload matters as much as overload because it reduces the ability to stay in control. The various coping strategies can be seen as typical or preferred ways of responding and not just as determined by the working conditions. The choice of a coping strategy not only represents a short term or temporary adjustment but may equally well indicate a more permanent style of work or management.

Complexity as a sleight-of-hand?

If the real purpose of complexity is to avoid that explanations become (too) complicated (something is complicated if it involves many different parts, in a way that is difficult to understand), it is a sleight-of-hand that just makes the epistemology simpler on the assumption that complexity is a true ontological concept.

Humans have a dislike of anything that is so complicated that it is difficult. We like simple terms and simple explanations. We like simple relations – preferably binary ones, either this or that, us or them, good or bad, safe or unsafe – or perhaps a line drawn in the sand. We like simple explanations and simple ways of understanding, such as the traditional cause-effect relations. And we like linearity or proportionality/symmetry, in the sense that outcomes are in proportion to causes – regardless of whether they are large or small – and also a symmetry in valence so that the cause of an unacceptable outcome itself must be something unacceptable, such as an error, an unwanted outcome, an injury cannot be caused by a trifle, such as performance variability. But when we are faced with systems or situations that are non-trivial and therefore irregular and partly unpredictable, these attractive symmetries and simplifications must be left behind.

The short-term solution to these problems is complexity. The term or the concept of complexity in one stroke allows us to maintain the efficiency or simplicity of explanations but still avoid the conceptual deprivation that is a consequence of the classical concepts and terms, such as causality and linearity. The problem is, however, that the nature of this complexity is epistemological rather than ontological. In other words, even though the descriptions have become simpler we are none the wiser about what they describe, if indeed it is possible to comprehend that at all. We must accept the fact that we are unable to go beyond or above the level of epistemological knowledge, and it is on that level that we need to make our explanations and our understanding clear. As a term, *complexity* appears to solve the problem, but it is a placebo rather than a proper solution. Complexity and complexity theory are forms of metaphysics, about which the eminent American philosopher Charles Sanders Peirce (2014, org.1878 p. 101) gave the following sound advice:

> Metaphysics is a subject much more curious than useful, the knowledge of which, like that of a sunken reef, serves chiefly to enable us to keep clear of it.

In von Förster's words the problem can be described as a change from "simple products by simple processes" to "complicated or non-trivial products by non-trivial processes." The products themselves are no longer trivial since they rarely are stable physical products in the traditional sense but services that have to correspond to the ever changing (unstable) reality.

> In part, the complexity sciences developed to research the behaviour of phenomena characterized by large numbers – and to use the computing power emergent from the 1980s onward.
>
> (Urry, 2005, p. 3)

Yet Urry continues:

> Following a deterministic set of rules, unpredictable yet patterned results can be generated, with small causes on occasions producing large effects and vice versa. The classic butterfly effect, accidentally discovered by Lorenz in 1961, demonstrated that minuscule changes at one location can theoretically produce, if modeled by three coupled non-linear equations, very large weather effects very far away in time and/or space from the original site of the hypothetical flapping wings.
>
> (Urry, 2005, p. 4)

(The butterfly effect is the idea that small things can have non-linear impacts on a complex system. The concept is famously illustrated by with a butterfly in Brazil flapping its wings and causing a tornado in Kansas.)

Yet similar complications may be found even in the three-body problem known from classical mechanics, first described by Isaac Newton in 1687. So complexity does not require fundamental unpredictability, either because the underlying system is stochastic (like throwing a die, stochastic but hardly complex) or because the system in question is intractable. Neither is it limited to phenomena characterised by large numbers. Three bodies will do.

The acceptance of non-linearity is: "The development of chaos theory involved rejecting the common-sense notion that only large changes in causes can produce large changes in effects" (Urry, 2005, p. 4).

Complexity investigates emergent, dynamic, and self-organising systems that interact in ways that heavily influence the probabilities of later events, hence being non-trivial as von Förster described it. Systems are irreducible to elementary laws or simple processes.

That is very well. But complexity is then a label for an approach or a science but not an explanation. So we have: complex in the ontological sense; complexity theory as a scientific approach; and the intractable systems that may or may not be complex. But they are definitely non-trivial.

> Central, then, to complexity is the idea of emergence. It is not that the sum is greater than the size of its parts – but that there are system effects that are different from their parts (see Jervis, 1997, on system effects). Complexity examines how components of a system through their interaction "spontaneously" develop collective properties or patterns, even simple properties such as color that do not seem implicit, or at least not implicit in the same way, within individual components (Nicolis, 1995). These are nonlinear consequences that are non-reducible to the very many individual components that comprise such activities.
>
> (Urry, 2005, p. 5).

Complex may be mistaken to be a property of the system (and therefore ontological); non-trivial makes clear that complexity is a property of the observer (and therefore epistemic). Systems are complex because what happens depends on previous operations (they have a history or past conditions), because they are analytically indeterminable and thus unpredictable. The dependence can be established as a fact and is in itself not complex, but the resulting behaviour or performance is.

It has already been described how the working environment and the industries where safety is a primary concern have changed dramatically in the last decades of the 20th century – since the 1950s as a matter of fact – and are now so different from the environment at the beginning of the century that comparison is all but impossible. Yet the basic thinking still goes back to those years. It is therefore not surprising that this way of looking at the world, this way of describing and explaining it, is unable to make much sense of what is going on today. Or rather, by trying to make sense of the world using the simple concepts of linear causality and all that follows from that, it is no surprise that the world looks complex. But it is complex in the epistemological sense that the descriptions are complex. The systems are not complex in the ontological sense.

It might therefore be argued that the need to invoke complexity mostly is an artefact of the impotence of the descriptive tools that are provided by the safety legacy (in particular the models and methods that are used to explain and analyse accidents, not least the limitations due to linear causality). If instead we had different and more powerful conceptual tools, there would perhaps be no need to invoke complexity. The world may still be complicated (as defined earlier) but would no longer be complex.

It is more than a little ironic that the invocation of complexity theory to explain what cannot be explained by linear ways of thinking in itself reinforces one part of the linear way of thinking, namely decomposition or reduction to a more fundamental level. This is like a kind of hierarchy of concepts, and the assumption is that the bottom of this hierarchy must be concepts that comply with the notion of causality and linearity. Perhaps we like Leucippus and Democritus are looking for the complexity atom.

An alternative might be to try something entirely different and abandon the use of causality. This is what happens when we accept that effects or results can be emergent as well as resultant as Lewes suggested in 1877, so it can hardly be said to be a new and revolutionary proposal. In the case of emergent consequences, we no longer need causality. The phenomena become easier to understand (without resorting to deconstruction), and the need to invoke complexity theory, which in itself is an ill-defined conglomerate of various things, nicely disappears.

What remains is that the world that we look at is complicated but not necessarily complex – and certainly not complex in the ontological sense. In a sense, emergence becomes the new ontology, and from that basis we skip the aetiological issues of causality.

Even if we turn to science, where words are carefully defined and have very precise meanings, we find that an agreed definition of "complexity" is not available (Cilliers & Richardson, 2001, p. 8).

The most useful definition that we personally have found refers to the definition of a complex system, rather than complexity metrics, which is simply given as a system that is composed of a large number of entities that display a high level of nonlinear interactivity. (Ibid)

But we already know, from the three-body problem in classical mechanics, that large numbers are not necessary. And non-linear is perhaps just a euphemism for unknown?

Complexity is really only an issue if we need absolute precision or certainty. Complexity is relative, as is being complicated. There is no degree of complication or complexity apart from the description, and the question must therefore be which purpose the description serves.

De Laplace argued that uncertainty is the same as ignorance. If we do not know something, we cannot say anything with certainty, and the uncertainty is only made more tangible by expressing it in terms of bits of information, for example the information that is required to determine whether something is true or not.

The uncertainty may be a fundamental issue at the quantum level but not at the macroscopic level. In fact, the more macroscopic you get, the less the uncertainty is. This is because performance is guided by social forces and norms and is predictable on that level. An example is the efficiency-thoroughness trade-off (ETTO) principle (Hollnagel, 2009). And it is just the kind of predictability we are interested in both in managing safety and in managing safely.

In relation to safety, we cannot predict with certainty when a person is going to do something and how it is going to be done. But we can say with high certainty, after knowing about the person and the social environment in which he or she works, that they are likely to do certain things, adjusting performance to the perceived conditions of making an efficiency-thoroughness trade-off. And we can also deduce or infer what the consequences of this can be – not with certainty but with reasonable assuredness for it to be of practical use, and easily enough to be able to think about what can be done to manage or dampen the variability.

So we can say something about the variability, the aggregate actions. This is just like in physics, where we do not need to know about the single electron to say how a current will flow. It is only if the demand for precision becomes too high that we run into problems.

Consider weather forecasts. Here it turns out that achieving more precise forecasts means that we must have a better understanding of the weather (or, similarly, of the economy). When we try to achieve that, we find out that there are more conditions that affect how the weather develops, hence that things become even more complicated – or even complex? So to get the desired prediction in weather forecasts we do need to know about more details, and things do get more complicated.

That is also the case for a lot of other phenomena, but not necessarily a problem for safety. We only want to know how the performance develops in general, or rather we need to get a feel for how people (individually or collectively) are likely to respond in certain conditions so that we can use these (estimated) responses as predictions of what will happen, for instance in planning the evacuation from a building (nightclubs apparently excepted), a subway station, a road tunnel, a large passenger vessel, or an aircraft and be able to decide whether it is sufficiently safe, meaning whether the evacuation is likely to go well or not, rather than whether it will fail or not. There is a limit to the precision we can obtain, and there is also a limit to the precision that we need. This is not so much due to complications, as to efficiency-thoroughness trade-off. Even if we knew more, we would not have the time and means to use that knowledge, lest safety managers become replaced by supercomputers. We cannot do a moment-to-moment control or management of work. In fact, that would go counter to the idea of Safety-II and deny the value of performance adjustments and the constructive role of people at both the sharp and the blunt ends, as emphasised by the so-called new view (Le Coze, 2022). These adjustments represent an irreducible uncertainty, and there is no need to go beyond that (except, perhaps, for academic psychologists). And since efficiency-thoroughness trade-off works on the individual as well as the collective and organisational levels, we do have a principle that is strong enough to give us the ability to analyse, manage, and improve safety – whether it is as Safety-II or as Safety-I.

The reasons why people prefer simple explanations are of course psychological (simple minds equals simple cognition). They are also social; it is more difficult (cognitively also) to explain something that is complicated, and it also requires more time and effort, precisely what we usually have too little of.

The final obstacle is cost. It takes more time and effort, and that takes resources away from doing things in the short run. But we forget that it is necessary to make a thoroughness-efficiency (TETO) trade-off in the short run to be able to make efficiency-thoroughness (ETTO) trade-offs in the long run. So we are always cursed by this dilemma. There is no formal or rational solution to this. It is a matter of policy and a recognition of what the mission of the company or organisation is. It is resilience thinking again, whether we should prioritise response (produce) and monitoring over learning and anticipation.

Complexity is only a problem if explanations rely on reductionism or decomposition, for example to be explained in terms of something else. A phenomenon as it is may be perplexing and possibly incomprehensible (when it is first encountered) and possibly confusing, perhaps even complicated – but it is never complex. The perplexity and confusion may exist to begin with but slowly (or quickly) disappear.

With regard to safety, we should not seek explanations by reduction and decomposition, as if the phenomenon in itself was not of sufficient value or *valeur*. We should seek explanations of the phenomenon itself, as a phenomenon. There may clearly be decisive factors and conditions that influence what is being done, but they do not explain why in any reductionistic sense. There may even be causality, either statistically or pragmatically. But there is never complexity.

In conclusion, complexity is a sleight-of-hand because the epistemology becomes simpler on the assumption that complexity is a true ontological concept.

The real purpose of complexity is to avoid that explanations become (too) complicated. Instead of having very detailed, intricate, and complicated explanations where the description of the system uses many terms and tries to account for many relations to make it predictable, the simple solution is to call it complex. Only in that sense can there be a single (monolithic) concept or cause, but it unfortunately does not enable the predictability that is necessary for effective control.

The attraction of linearity

The attractive feature of linearity is that if you have *n* elements or steps, then there are only *n-1* connections or perhaps a few more if branching is allowed; the system is therefore in principle predictable even for the natural human brain. But if there are non-linear or non-sequential relationships, where in principle everything can be connected to everything else, this may easily lead to *n!* connections. And since our brains only allow us to reason sequentially, at least for the common man, and because since Miller (1956) it has been widely accepted that short-term memory is limited so that seven or so items can be considered simultaneously, then the conclusion is obvious. By describing systems as linear we can consider seven elements and perhaps a few more if we chunk. By describing systems as non-linear, we are down to three elements, since $3! = 6$ and $4! = 24$.

Humans have an innate dislike of anything that is so complicated that it is difficult, whatever the explanation may be. We like simple terms and simple explanations. We like simple relations that are predictable, preferably binary ones, either this or that. We like simple explanations and simple ways of understanding, such as in the cause-effect relations. And we like linearity or proportionality, in the sense that outcomes that are in proportion to causes is predictable – regardless of whether they are large or small.

But when we are faced with systems or situations that are non-trivial – in the sense that they are irregular and partly unpredictable, in the sense that the descriptions are complicated with many constituent concepts and relations, and in the sense that there are phenomena happening that we do not fully understand (making it almost a recursive affair) – then we must surrender ease of explanations and ease of understanding. We have to give up efficiency – meaning simplicity – for thoroughness, since we do need a degree of understanding that allows us to manage the phenomena. (If we did not have to manage them but could submit to fate or follow the will of the gods, it would be easy enough to find explanations and descriptions that were overly simplified, but it would not be a good way to run a commercial company.)

The way out of this problem is complexity. The term – or the concept – *complexity* in one stroke allows us to maintain the efficiency or simplicity of the descriptions but still to avoid the over-simplification that is an unavoidable consequence of the legacy of concepts and terms, such as causality and linearity. The problem is, however, that complexity is on the level of epistemology rather than ontology. In other words, we are none the wiser about what actually goes on, if indeed it is possible to understand that at all. We must accept the fact that we remain on the level of epistemological knowledge, and it is on that level that we need to make our explanations and our understanding clear. Complexity appears to solve the problem, but it is a placebo rather than a true medicine.

Complexity as a metaphorical workhorse

The intention here is not to disagree that there is a complexity science, since there probably is (whatever a science may be) and it may even be useful in many cases. The intention is rather to argue that the problems of complexity are not solved by calling systems complex. The issue is not whether complexity is ontological or epistemic; it is clearly the latter rather than the former, which makes the problem both easier and more difficult to solve. It makes it easier because if complexity was ontological, for instance if the socio-technical systems we struggle with really were complex, then the only solution would be to make them less complex, but that would require that we could describe them well enough to know what to do, which would mean solving the problem of epistemological complexity potential ontological complexity first, but that must be solved anyway. The essential question is therefore: how can we simplify the descriptions of complexity?

What is complexity?

According to Urry (2005) complexity sciences developed to research the behaviour of phenomena characterised by large numbers – and to use the computing power emergent from the 1980s onward (p. 3).

Complexity also has to do with non-linearity: "The development of chaos theory involved rejecting the common-sense notion that only large changes in causes can produce large changes in effects" (p. 4).

Complexity investigates emergent, dynamic, and self-organising systems that interact in ways that heavily influence the probabilities of later events. Systems are irreducible to elementary laws or simple processes.

That is very well. But complexity is then a label for an approach or a science and not an explanation. So we have: complex in the ontological sense; complexity theory as a scientific approach; and the non-trivial systems that may or may not be complex. But they are non-trivial.

If we compare the accident models that are part of Safety-I, which means models that are linear and de-compositional, then it is not surprising that they are unable to account for the modern world. As described in Part I, these models have their root in the thinking of the early 20th century and indeed go back for at least a couple of thousand years – to the extent that anyone tried to think systematically about safety then (and we have no written evidence to confirm that. The possible exception is the code of Hammurabi which clearly embodies the principle of causality and therefore responsibility. But this was legislation and not a scientific theory. The first model that is formally described is the domino model, which clearly is linear causality embodied. What came after is basically embellishments, although often very elegant ones, but they never let go of the assumption that both causality and linearity were valid principles. Even a relatively new accident model such as system-theoretic accident model and processes (STAMP) cannot break free of the safety legacy's cognitive straitjacket, as this description by Leveson (2016) makes abundantly clear.

> STAMP Examines parts separately and later combines analysis results
> The STAMP model therefore assumes such separation does not distort phenomenon
> It further assumes that each component or subsystem operates independently

And also that Components act the same when examined singly as when, playing their part in the whole. This is practically an endorsement of the substitution myth (Bradshaw et al., 2013).

And finally that events not subject to feedback loops and non-linear interactions. STAMP therefore requires that the systems being analysed and modeled are trivial according to von Förster's terminology.

Part II has also described how the working environment and the industries where safety is a primary concern have changed dramatically in the last decades of the 20th century – mostly since the 1950s as a matter of fact – and is now so different from the environment at the beginning of the century that comparison is all but impossible. Yet the basic thinking embodied in the safety legacy is still the old one. It is therefore not surprising that this way of looking at the world, this way of describing and explaining it, is incapable of making much sense of what is going on. Or rather, by trying to make sense of the world using the simple concepts of linear causality and all that follows from that, it is no surprise that the world looks complex. But it is complex in the epistemological sense, that the descriptions are complex. It is not complex in the ontological sense, that it really *is* complex. There is, anyway, no way we can know that apart from the descriptions, which sort of takes us back to where we started.

It can thus be argued that the need to invoke complexity is an artefact of the impotence of our descriptive tools (the accident models, in particular the assumption of causality). If instead we have different and more powerful conceptual tools, there might be no need to invoke complexity. The world might still be complicated but it would no longer be complex.

An alternative solution is to abandon the role of causality, for instance, by acknowledging that effects or results can be emergent as well as resultant. In the case of emergent consequences, we no longer need causality. The phenomena become easier to understand, and the need to invoke complexity theory, which in itself is an ill-defined conglomerate of various things, would simply disappear.

What remains is that the world that we look at is complicated but not necessarily complex – and certainly not complex in the ontological sense. In a sense, emergence becomes the new ontology, and from that basis we skip the aetiological issues of causality.

Even if we turn to science, where words are carefully defined and have very precise meanings, we find that an agreed definition of "complexity" is not available (Cilliers & Richardson, 2001, p. 8).

The most useful definition that we personally have found refers to the definition of a complex system, rather than complexity metrics, which is simply given as a system that is comprised of a large number of entities that display a high level of nonlinear interactivity (Ibid).

But we already know that the large number is not necessary. And is *non-linear* not just a euphemism for *hard to describe*?

Complexity is really only an issue if we need absolute precision or certainty. Complexity is relative, as is being complicated. There is not degree of complication or complexity apart from the description, and the question must therefore be which purpose the description serves.

Laplace pointed out that uncertainty basically is the same as ignorance. If we do not know something, we cannot say anything about it with certainty, where the uncertainty is made more tangible by expressing it in terms of bits of information, for example, the information that is required to determine whether something is true or not.

The uncertainty may be a fundamental issue at the quantum level but not at the macroscopic level. In fact, the more macroscopic you get, the less the uncertainty is. Take the example of people eating every night. One of my colleagues liked to point out that it is quite easy to predict what people in the industrialised part of the world will do at about 6 pm. They will have dinner in their homes, unless their country is ravaged by a war. The precision or certainty increases if you can narrow the subculture or the sample like the people in a suburb of Stockholm– but only to a certain level. If you get to the individual person, the certainty disappears again. This is because performance on one level is guided by social forces and norms and therefore is predictable on that level, as in the ETTO principle. And it is just that kind of predictability we are interested in.

In relation to safety, we cannot predict with certainty when a person is going to do something and how it is going to be done. But we can say with high certainty, after knowing about the person and the social environment in which he or she works, that they are likely to do certain things, to make some habitual adjustments, both because it seems to be the right thing to do and more importantly because it has worked in previous cases, and their colleagues expect them to do so. Calling this non-compliance or a failure mode is therefore not of as much help as ETTOing. And we can also deduce or infer what the consequences of this can be. Not with certainty but with reasonable assuredness. Enough to be able to think about what can be done to manage or dampen the variability.

So we can say something about the variability, the aggregate actions. This is just like in physics, where we do not need to know about the single electron to say how a current will flow. It is only if the demand to precision becomes too high that we run into problems.

Consider weather forecasts. Here it turns out that achieving more precise forecasts means that we must have a better understanding of the weather (or similarly of the economy). When we try to achieve that, we find out that there are more conditions that affect how the weather or the economy develops, hence things become more complicated– or even complex? So to get the desired prediction in weather forecasts we do need to know more details, and things do get more complicated.

That is also the case for many other phenomena. But this is not necessarily a problem for safety. We only want to know how the performance develops in general, or rather we need to get a feel for how people (collectively or individually) are likely to respond in certain conditions, so that we can use these (estimated) responses as predictions of what will happen and be able to decide whether it is sufficiently safe or not – whether it will succeed or not, rather than whether it will fail or not. There is a limit to the precision we can obtain and also a limit to the precision that we need. This is not so much due to complications as to the effects of the ETTO principle. Even if we knew more, we would not have the time and means to use that knowledge. We cannot do a moment-to-moment control or management of work. In fact, that would go counter to the idea of Safety-II and deny the value of performance adjustments. These adjustments represent the irreducible uncertainty, and there is no need to go beyond that (except for academic psychologists, perhaps). And since efficiency-thoroughness trade-off work on the individual as well as the collective and organisational levels, we do have a principle that is strong enough to give us the ability to analyse, manage, and improve safety – as Safety-II as well as Safety-I.

The final obstacle is cost. It takes more time and effort, and that takes resources away from doing things in the short run. But we forget that we must make a trade off efficiency for thoroughness-efficiency trade-off (TETO) in the short run to be able to trade off thoroughness for efficiency (make efficiency-thoroughness trade-offs) in the long run; we can only be efficient in the present if we have prepared for that by being thorough in the past. The dilemma is inescapable. There is no formal or rational solution; it is a matter of policy and a recognition of what the mission of the company or organisation is. In resilience engineering terms it is the question of whether we should prioritise responding (production) and monitoring over learning and anticipation.

Complexity is only a problem if explanations must rely on reductionism or decomposition, for example, being explained in terms of something else. A phenomenon as it is may be perplexing and possibly incomprehensible (in the moment) and even confusing, perhaps even complicated – but it is never complex. The perplexity and confusion may exist to begin with but slowly (or quickly) disappear.

With regard to safety, we should not seek explanations by reduction and decomposition, as if the phenomenon in itself was not of sufficient value or *valeur*. We should seek explanations of the phenomenon itself, as a phenomenon. There may clearly be decisive factors and conditions that influence what is being done, but they can never explain it in a reductionistic sense. There may even be causality, either statistically or pragmatically. But there is never complexity.

The complexity paradox

That complexity paradox has unintentionally been defined by (Cilliers, 2005). First Cilliers wrote that "if we acknowledge that the world in which we have to live is complex; we also have to acknowledge the limitations of our understanding of this world" (p. 256). Two pages later Cilliers wrote that "To fully understand a complex system, we need to understand it in all its complexity" (p. 258). In other words, we have to do what we cannot do. The complexity paradox is not in itself complex but nevertheless it is impossible to solve.

Cillier's dilemma bring to mind the famous seventh statement by which Ludwig Wittgenstein (1922) ended his *Tractatus-Logicus-Philosophicus*: "What we cannot speak about we must pass over in silence." According to Cilliers' (2005) definitions, complexity is not just something about which we cannot speak; it is also something about which we cannot think! It therefore seems appropriate to end Part II here. I, of course, apologise to the reader for not having made this clear from the beginning.

Complexity epilogue

Part II has discussed whether complexity is epistemic or ontological and considered the practical consequences of either position both for safety management in the conventional interpretation and for managing safely. Complexity is a problem regardless of whether it is epistemic or ontological. If complexity is epistemic then it is necessary to develop new and more powerful ways of describing systems and their dynamics. If complexity is ontological, then there is no direct solution. Either way, safety management or managing safely must overcome two major obstacles in order to move forward. The first is the impotence of the conceptual and methodological legacy described in Part I. The other is the problems due to the non-trivial nature of contemporary industrial environments, described in Part II. The remaining parts will do their best to explain how both obstacles may be overcome.

III The futility of accident investigation

Futility prologue

Part III continues the critical analysis from Part I of accident investigation as an approach to learn about and improve safety and argues that the limitations of the safety legacy are incompatible with the non-trivial nature of present-day societies and activities, as Charles Perrow had pointed out already in 1984. Current accident analysis methods are then briefly analysed to highlight their built-in assumptions. The need for certainty is partly driven by the Heinrich dogma and the *safety mantra*, the soothing phrases that executives and politicians utter in the aftermath of a major accident – often with little correspondence to the accident itself or the severity of the outcomes – and the prescriptions thereof. This leads into a discussion about the nature of safety, insofar as this is a meaningful concept in the first place. It is also pointed out how the recommendations that follow the analyses are both driven and constrained by the underlying thinking, and they therefore often are more similar than the accidents they are assumed to address. This is also due to the fact that causes are social constructs that do not always have a physical reality.

Safety and learning

Part III begins by looking at the relation among reporting, learning, and safety from a Safety-II perspective. It is argued that reporting should serve the purposes of learning and hence be a precondition for learning, rather than the other way around. The prevailing practices, following the Heinrich dogma, are to analyse whatever has been reported and then try to learn from that. In this case learning becomes a consequence of reporting rather than a precondition to it. Since accidents are stochastic there is no control over which events happened and were reported and therefore no control over *when* learning takes place. Events are simply considered relevant for safety and for learning by virtue of having been reported. This makes it impossible to support learning as a continuous effort. Since Safety-II is defined as a condition where as much as possible goes well, learning will here focus on what should be done rather than on what should not be done, hence representing a positive rather than a negative stance. (These issues are also discussed in Part V Coda).

DOI: 10.4324/9781032664729-3

The accident pyramid

The accident pyramid, which because it is so widely known, basically shows a pyramid with a number of horizontal layers, each layer representing a certain category of outcomes together with the frequency or number of occurrences of outcomes in that category. In the Heinrich book (Heinrich, 1959, p. 29 and Figure III.1) the bottom layer was called "no injury accidents" and the corresponding number was 300; the next layer was called "minor injuries" and the number was 29 (see Figure III.1). The third layer was called "major injury," and the number was just 1. But there are many other versions of the accident pyramid in both 2D or 3D and multiple colours. In one the layers and numbers are as follows:10,000 unsafe acts and conditions, 1,000 near misses, 100 cases where medical attention is needed, 10 loss time injuries, and at the top 1 fatality.

The accident pyramid, sometimes also called the Bird triangle (after Frank E. Bird who was director of engineering services for Insurance Co. of North America), implies a theory of industrial accident prevention and proposes a relationship among serious accidents, minor accidents, and near misses. The accident pyramid gives the impression that if the number of minor accidents is reduced then there will be a corresponding fall in the number of more serious outcomes.

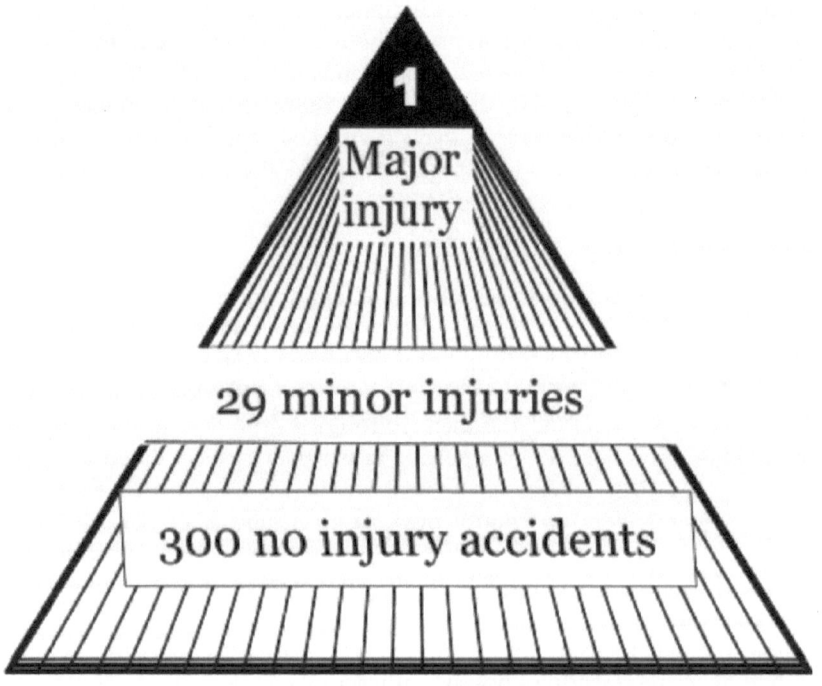

Figure III.1 The original accident pyramid (based on Heinrich, 1959, p. 29)

Table III.1 The contents of three different accident pyramids (from Hale, 2002, p. 35)

Source						
Source	*Heinrich (1931)*		*Bird's triangle*		*Saloniemi's pyramid*	
	Outcome type	number	Outcome type	number	Outcome type	number
	No injury	300	Property damage	500	Near accident	105,371
	Minor injury	100	Minor injury	100	Minor injury	8,923
	Disabling injury	1	Disabling injury	1	Fatal accident	88

This is actually stated explicitly in the text of the figure that appears in Heinrich (1959) and which has been included here as Figure III.1. Here called the "moral" (sic!), Hale's (2002) paper on "Conditions of occurrence of major and minor accidents Urban myths, deviations and accident scenarios" included a figure with three pyramids with different numbers The figures are not reproduced here but the contents are listed in Table III.1.

Pyramid problems

There are unfortunately several problems with the accident pyramid. The categories are first of all neither unique nor well defined (see Table III.1). And if the categories are not well defined, how is it then possible to report or count the number of occurrences of each? And can the numbers actually be trusted? A second and worse problem is that the causal relation between the layers is suggested but never articulated. There is no strong theory – and not even a weak theory to support it.

A third problem is that the iconic pyramid is based on a visual misinterpretation of the illustration in one of the Heinrich works, reproduced in Figure III.1. It is easy to see that Figure III.1 tries to show the various categories in a perspective, and is not a pyramid at all, although the misinterpretation is understandable. The accident pyramid is also based on a conceptual misinterpretation. Heinrich's own caption for the illustration was "The foundation of a major injury." Heinrich (1931) had also introduced the so-called iceberg model to highlight the hidden costs of accidents; it refers to the analogy of an iceberg, although there is no explicit reference to an iceberg where only the tip is visible above the water, but where nine tenths of the iceberg is below the surface and hence cannot be seen. There is no illustration of an iceberg either in Heinrich (1931) or in Heinrich (1959) and no direct mentioning of icebergs, just a reference to the hidden costs of accidents. An iceberg is a powerful and widely used analogy. When rendered graphically, it can, however easily be misunderstood because the triangular contour of an iceberg resembles a pyramid. The whole idea of an accident pyramid is therefore based on a coincidence of multiple confusions or misunderstanding, as an interesting parallel to how accidents happen. But even if the accident pyramid had been correct, there are no good reasons to assume that there is more to learn by analysing the serious outcomes, rather than the less serious. The only real advantage is that it must be done less often. Some companies are known to take the pyramid very literally and think that as long as their reported incident numbers correspond roughly to the ratios found in one of the many versions of the pyramid, then there is nothing to worry about.

A final problem is that the different proportions of the various types of outcomes do, however, not imply or prove causality. Another and simpler explanation in line with contemporary thinking is that the more serious outcomes depend on a larger number of failures and conditions coinciding, and this in itself explains that they are less probable, hence that their number is lower. This is just as in the game of poker where getting five of a kind is very unlikely ($p = 0.000000384$). Perhaps Heinrich did not play poker?

Heinrich's original figure included the following text:

0.03 % of all accidents produce major injuries
0.88 % of all accidents produce minor injuries
90.9 % of all accidents produce no injuries

The ratios graphically portrayed in Figure III.1 show that in a unit group of 330 similar accidents (but similar in which way?), 300 will produce no injury whatever (this raises the interesting question of how Heinrich or anyone else, could know there had been an accident if there had been no injury (people report injuries to insurance companies, but not accidents.) Twenty-nine will result only in minor injuries and 1 will result seriously. The major injury may result from the very first accident or from any other accident in the group.

MORAL – PREVENT THE ACCIDENTS AND THE INJURIES WILL TAKE CARE OF THEMSELVES. (Capital letters as in the original).

The accident pyramid was first proposed by Herbert William Heinrich in 1931 (but did not appear until the 1959 version of his book). The pyramid has been endlessly copied ever since (Hale, 2002, p. 35). The accident pyramid involves an unasserted assumption of simple linear causality, much as the domino model by the same author did.

Hale (2002, p. 35) provided the following comment to Heinrich's idea of an accident pyramid

Heinrich was content to draw the rather broad conclusion that the seriousness of the consequences of an accident had a strong random elements and that minor accidents often preceded major accidents. Examples quoted in his book shows how loose this reasoning is for him; the minor injuries cited take place in the same activity but have a different and only partially over-lapping set of causes from the major injury described, e.g. twisting an ankle tripping over the rail tracks when taking a short cut into the works vs. get-ting crushed to death between two trucks shunted on those tracks. Hence we cannot check his reasoning in detail. However, the example which is often quoted later to prove the point is of an accident caused by an object falling from a crane, which may hit and kill, graze and slightly injure, or miss a person standing underneath. This anecdote is instantly recognisable and seems to prove the principle of similar cause for major and minor incidents. Heinrich himself takes the argument further in quoting the result of his later study into accident prevention in 100 manufacturing plants (Heinrich et al., 1980).

"The causes of serious injury accidents did not fairly picture the unsafe practices and conditions needing attention. Accident-prevention work in these plants was misdirected, since it was based upon the investigation of major injuries, and many other serious injuries of a slightly different nature later occurred." And further: The conclusions drawn from the pyramids were that "The causes of serious injury accidents did not fairly picture the unsafe practices and conditions needing attention. Accident-prevention work in these plants was misdirected, since it was based upon the investiga-tion of major injuries, and many other serious injuries of a slightly different nature later occurred."

This was taken by his followers as showing that attention to minor accidents would be more successful in directing prevention. However, that goes far beyond the evidence, and note also that Heinrich himself does not mention minor accidents, only near misses. He only claims that removing the causes of past serious injuries does not prevent future serious injuries. This may have been true in his case, but may only show that a small sample of major accidents is not representative of the total possible. One can ask whether his time scale was long enough to demonstrate his point. "Studies done over the past ten years at a steelworks Swuste et al. (2002a, 2002b) indicate that such a reactive policy based almost exclusively on lost time accidents can produce good results: a reduction in the lost time accident rate of 30 percent and of total accident rate of 60 percent in nine years, not dramatic but a steady improvement. It is, however, clear that such an approach will only work in a relatively stable workplace and technology, which is not introducing new hazards faster than they are being removed by this incremental prevention." (Hale, 2002, pp. 35–36).

What is surprising is the strength of the belief in identical causes of major and minor accidents which, subsequent to Heinrich's original work, grew up among safety practitioners and apparently also among researchers. The persistency of this belief is shown by the vigour with which the dissenting voices have had to express themselves to counter this belief (Petersen, 1971, 1989; Hale & Hale,

1972; Saloniemi et al., 1992; Saloniemi & Oksanen, 1998). Even the fifth edition of Heinrich's book found it necessary to state that:

> There has been much confusion about the original ratio in industrial acci-
> dent prevention. It does not mean, as we have too often interpreted it to
> mean, that the causes of frequency are the same as the causes of severe
> injury. National figures show that different things cause severe injuries than
> the things that cause minor injuries. Statistics show that we have been only
> partially successful in reducing severity by attacking frequency.
>
> (Heinrich et al., 1980)

This belief seems to be an example of an urban myth: a belief which seems so plausible that it commands immediate acceptance without proof. Clearly such a belief
is so convenient for safety practitioners that they hardly stop to question it. The
strength of the belief can be understood from the viewpoint of a company where
minor accidents occur relatively frequency and major accidents are rare. They see
their efforts and success in reducing minor accidents. They see no major accident.
The temptation is to draw a cause-effect link, without questioning whether there
would have been major accidents even without the efforts in preventing minor
accidents. "What is then the direct research evidence?" (Hale, 2002, p. 36).

The problems with causes

Looking for causes

(Much of the following text has previously appeared in Chapter 1 of (Hollnagel,
2004a)). Whenever an accident happens there is a natural concern to find out in
detail exactly what happened and to determine the causes of it. Indeed, whenever
the outcome of an action or event falls significantly short of what was expected
or whenever something unexpected happens, people try to find an explanation
for it. This trait of human nature is so strong that we try to find causes even when
we ought to know they do not exist, such as in the case of misleading or spurious
correlations. For a number of reasons humans seem to be extremely reluctant
to accept that something can happen by chance (Landsman & Van Wolde, 2016).
One very good reason is that we have created a way of living that depends heavily on the use of technology, which is constructed to function in a deterministic,
hence reliable, manner. If therefore something fails, we are fully justified in trying
to find the cause for it. A second reason is that our whole understanding of the
world is based on the assumption of specific relations between causes and effects,
as amply illustrated by the laws of physics. (Even in quantum physics there are
assumptions of more fundamental relations that are deterministic.) A third reason
is that most humans find it very uncomfortable when they do not know what
to expect, i.e., when things happen in an unpredictable manner. This creates a
sense of being out of control, something that is never desirable since – from an

evolutionary perspective – it means that the chances of survival are reduced. The condition was perfectly described by the great German philosopher Friedrich Nietzsche (1844–1900) when he wrote that:

> to trace something unknown back to something known is alleviating, soothing, gratifying and gives moreover a feeling of power. Danger, disquiet, anxiety attend the unknown – the first instinct is to eliminate these distressing states. First principle: any explanation is better than none. . . . The cause creating drive is thus conditioned and excited by the feeling of fear.
>
> (Nietzsche 1977, p. 62)

A well-known example of the search for a causal relation is provided by the phenomenon called the *gambler's fallacy*. The name refers to the fact that gamblers often seem to believe that an extended series of outcome of one type increases the probability of the complementary outcome. Thus if a series of "red" outcomes occur on a roulette wheel, the gambler's fallacy leads people to believe that the probability of "black" increases or, conversely, that one positive outcome makes it less likely that another one will follow soon. Rather than accepting that the underlying mechanism may be random, people invent all kinds of explanations to reduce the uncertainty of future events. The general tendency to attribute rare events to causal factors is furthermore not the prerogative of gamblers but may be found even in scientific-sounding statements such as "Given all the logically possible combinations of which DNA molecules are capable, the odds against human life evolving are staggering. The fact that we are here proves that some intelligence guided the process to make our appearance inevitable." As Nietzsche said, any explanation is better than none.

Facts and explanations

This part of the book is not about error, neither the dreaded individual "human error" nor organisational errors. The basic reason is that a focus on errors takes for granted that this is the most important thing to look at. It also implies that a simple cause-effect model is the only reasonable one, instead of keeping an open mind. There are, in fact several ways in which an accident can be investigated and understood, and although the linear cause-effect assumption is easy to use it is perhaps the least attractive option. A further reason is that a number of works have already been published that go into discussions of errors at great length, although often with widely diverging views and opinions. Examples are Reason (1990a, 1997), Senders and Moray (1991), Rasmussen et al. (1987), Hollnagel (1993, 1998), Woods et al. (1994), and many others. So rather than attempting yet another analysis of error (Hollnagel, 2004a, p. 26).

The investigation of events with unacceptable outcomes (aka accidents) – and the search for explanations – is often based on the assumption, incorrect as it turns out, that explanations can be deduced from the facts. Thus, accident investigations and the search for causes are very often just trying to fit all the facts together, in the belief that there is some kind of objective truth just waiting to be found. Investigating an accident is treated as if it was a detective story where the great detective tries to fit all the facts together. But we can first of all never be sure that we have all the facts; something may well be missing. Also, some data and information may not represent facts but rather be spurious observations that are temporally contiguous but otherwise causally unrelated. Finally, the facts such as they are are not independent of the accident model and the method that goes with it. Indeed, facts are not found but sought out, as described in Part I.

The difference between explanations and causes

An accident investigation is an attempt to find out both *how* the accident happened and *why* it happened. An investigation should be necessarily systematic, so that the resulting account of what happened is biased neither by premature assumptions nor pet hypotheses, nor should it invoke ad hoc psychological or organisational causes or counterfactual conditions, which often only substitute one term for another. To ensure that, human factors researchers and safety-conscious practitioners have over the years developed a set of accident analysis methods and classification systems. All methods necessarily imply an accident model, i.e., a mutually agreed and often unspoken understanding of how accidents occur. While there have been several changes to our understanding of accidents over the last couple of decades, such changes have mostly addressed the nature of the causes, as described by Hale and Hovden, 1998. We have gone from a focus on technological deficiencies, to a focus on the negative impact of human performance failures ("human error"), to a focus on the role of organisational factors. Yet all of these explanations rely on the same tacit accident model, which usually is taken for granted and is shared by the people involved in the analysis.

The tendency to look for causes rather than explanations is often reinforced by the methods that are used for accident analysis. The most obvious example of that is the principle of root cause analysis (RCA), for example Cojazzi and Pinola (1994), which is commonly applied in many different domains. As the name implies, the principle entails that it is possible to find a basic cause that is the root or origin of the problems, specifically of the incidents and accidents that occur. It is thus an expression of a strong principle of causality combined with simple linear reasoning, derived from the Heinrich dogma and the Domino model.

Most root cause analysis methods propose a number of rules for how the analysis shall be carried out. One rule may be that causal statements must clearly show a "cause and effect" relationship. This advice is deceptively simple and seemingly innocent. But it hides the assumption that it is possible to establish a "cause and effect" relationship, which means: (1) that nothing happens without a cause *and* (2) that it is possible to find that cause from knowledge of the effect. It further implies, though more subtly, that if the cause is found and eliminated in some way, then the accident will not happen again. This is practically synony-mous with the Heinrich dogma, that "the occurrence of an injury invariably results from a completed sequence of factors – the last one of these being the accident itself." Both the rule and the axiom correspond to a simple model of an accident as a sequence of events, most vividly represented by the Domino model of accident causation.

What we can say is rather that the events we call accidents happen due to a number of factors coinciding or becoming aligned at a specific time but not that the accident was caused by any single one of them. An excellent example of that is the meltdown at the nuclear power plant in Harrisburg, Pennsylvania on March 28, 1979. Seen together, the several factors and conditions constitute an explanation, in the sense that we then can understand *how* the accident hap-pened. Yet this does not mean that the explanation or any of the factors or condi-tions it refers to are the cause, and it does not tell us *why* the accident happened or why the coincidence occurred. The cause, if any, is in the concurrence or coincidence of these various factors. Yet since it in any practical sense is impos-sible to eliminate coincidences (due to complexity, cf., Part II), it is also impos-sible to eliminate causes of accidents as such. The difference between looking for explanations and for causes is thus crucial. If accidents really did have causes, then it would make sense to try to find them and do something about them once found. If accidents merely have explanations, then we should rather try to account for how work took place and for what the conditions or events were that led to the unacceptable outcome, which requires that we also pay attention to the dynamic non-events when work goes well! The response should not be to seek out and destroy causes but to identify the conditions that may lead to an unacceptable outcome and find effective ways of controlling them. Indeed, according to Hale and Glendon, 1978, UK factory inspectors in the 19th cen-tury were only interested in getting reports of accidents with technical causes, since others could not reasonably be prevented. (An unusual, but refreshingly honest admission of their own limitations.)

From technological failure to "human error"

Throughout the history of accident investigation there has been a strong and natural tendency to look for explanations or causes in those parts of the systems, which fail most frequently or which in some ways are conspicuous. In the early phases, i.e., until the late 1950s, those were the clunky technological or mechani-cal parts of systems. People knew from their everyday experience that tech-nology on the whole was prone to fail or malfunction (Leveson, 1992). When therefore an accident – and in particular a spectacular accident – happened, there was a natural tendency to look for causes that could be expressed in terms of technological failures.

As time went by technology became more reliable and therefore receded into the background as the most likely cause of accidents. In its place came the human action – infamously known as "human error." Finding the explanation in human actions rather than in a technological malfunction is, of course, not enough, since that is a wholesale description almost on the level of acts of nature. Furthermore, since no system creates or maintains itself, the search for a human action is bound to succeed if it only goes on for long enough, which makes it a safe bet in the search for causes. When human action as such was identified as a cause, then little could be done except to blame or punish the person in some way. There was a clear need to go further, for instance to find an explanation beyond the explanation of "human error" (Woods & Cook, 2002). Here human factors and information processing psychology came to help because they provided a powerful (but wrong) analogy from which explanations could be constructed.

The development of human factors took a big step forward when psychologists and others in the late 1950s discovered the computer metaphor and realised that the human mind could be described as if it was an information processing system (which it definitely isn't). The computer metaphor made it possible to decompose human behaviour into a set of basic structures and associated processes in complete analogy with the description of technological systems (Rasmussen, 1974). Just as for technological systems, human information processing components could fail, if not directly break, and such failures could be seen as the cause of accidents through more-or-less well-defined cause-effect relations. The analogy was even taken as far as proposing the existence of a "human error mechanism" although this ill-advised practice later fell into well-deserved disrepute.

The early enthusiasm for "human error" as universal source of accident causes has now waned, one reason being that it became rather obvious that it was all too easy to invent new "error categories" to fit a situation. (It is indeed a convenient feature of the human mind that it is always possible to suggest a new intermediate function or link, hence the plethora of "human error" categories and cognitive functions.) And simply explaining one problem by substituting it with another may be great fun, but it is a futile exercise in the long run. In its place the focus was turned to the relation between organisational factors and accidents (Reason, 1997), although organisational factors in themselves at some level require an explanation on the level of the individual. Curiously enough, this development has been horizontal rather than vertical, in the sense that organisational factors have been seen as yet another link in the cause-effect relations, rather than as a transition to a qualitatively new level of explanation. The reason for that is probably that the underlying accident model has remained the same, namely the sequential accident model (Hollnagel, 2002). Yet it is clear, according to the previous line of argument, that the concept of "human error" is an artefact of a theoretical development coupled to a technological development. It is therefore time to reassess the term and hopefully take a step further.

Causality as an artefact of temporal order

One of the most important – and possibly most overlooked – relations in accident analysis is between causality and temporal order. As already discussed, it is a general axiom of causality that the cause must be prior to the effect – barring exotic fields such as quantum physics. There is, however, another consequence or effect of time, namely that events as we recall them are ordered in time.

In relation to accident analysis this means that any description of the events that occurred before the accident – regardless of whether they were relevant for the accident or not – imposes a sequence or order. This is a consequence of the simple fact that time progresses in one direction only and that we experience and remember events one at a time as happening one after the other, even when they are factually simultaneous. However, the fact that event A occurs before event B may be a *necessary* condition for A being a cause of B, but it is not a *sufficient* condition. This was pointed out already by Scottish philosopher David Hume (1711–1776), who in *Treatise of Human Nature*, published in 1739, provided a groundbreaking analysis of causality. Despite that, we have a deplorable tendency to equate temporal ordering with causality, particularly if the two events seem to have something in common, as nicely captured by the Latin phrase *post hoc ergo propter hoc*.

What is more and worse, is that we are misled by the fact that we can describe past events as a sequence to believe that we can also describe future events as a sequence. We cannot only describe past events as a sequence but even as a tree, using the fallacious counterfactual reasoning. In this way we can examine or speculate about hypothetical developments that did not take place, either the ones that would have been beneficial or better or the ones that would have been detrimental or worse (and hence praise ourselves and be thankful that they did not happen).

This possibility creates the incorrect belief that we can describe hypothetical future events in the same way. The reason why it is possible to describe past events in a sequence is that they *did* occur and that therefore the temporal relations between the events are manifest. Hypothetical events can also be positioned relative to the manifest ordering of the real events. The same is not the case for future events. It is, of course, possible to consider alternative ways of development for a single future event (common tools are fault trees and event trees even put together as a cause-consequence tree; Taylor, 1976) for instance in the Bowtie model (de Ruijter & Guldenmund, 2016), such as failure or error modes, but it is not reasonable to assume an immutable order of two, three, or more events in a sequence. Although it may indeed be reasonable to assume that in most cases event B will occur after event A – due to the laws of physics, for instance – the order is not permanently fixed. And it is precisely the unexpected reversals of order or breaking of the envisaged temporal bonds that often lead to serious unacceptable outcomes.

Evolving concepts of causes

Even though finding and eliminating the causes cannot always prevent accidents, it is still instructive to look for causes as parts of the explanations, rather than as root causes. It is also instructive to see how the categories of causes have developed through the years, as described by the three ages of safety (Hale & Hovden, 1998), reflecting a development from simple accident explanations to complex ones – from simple causality to complex coincidences (Figure III.2). This is, of course, not unrelated to the de facto growth in the complexity of technological systems.

Going back to the early days of industrial accident analysis, which means the first half of the 20th century stretching into the 1970s, the main categories of causes were technical failure, human error, and other. (This rough characterisation obviously does injustice to a number of people who were far ahead of their time. It is nevertheless a fair rendering of the views of the mainstream, particularly the views of the engineering community as a whole, i.e., the non-specialists.) The reason is not difficult to find, since technical systems on the whole were less reliable than today. As an illustration, just consider a car from the 1950s and one from 2000. Despite the considerably larger number of components in a modern car, it is not only better performing but also more reliable.

Over the years there have been periodic developments within each of the three main categories either because new failure types arose or because of developments in the corresponding theories and models. As far as "human error" is concerned, a major boost began in the mid-1970s when mainstream human factors specialists and psychologists adopted the human information processing view and used that both to enrich their understanding of the nature of human action and to produce a flourishing vocabulary. This development was initially focused on "human errors" during operation or actual work at the sharp end, but soon grew to include other categories of human work, specifically maintenance, management, and design. The considerable literature on "human error" mentioned earlier is testimony to that.

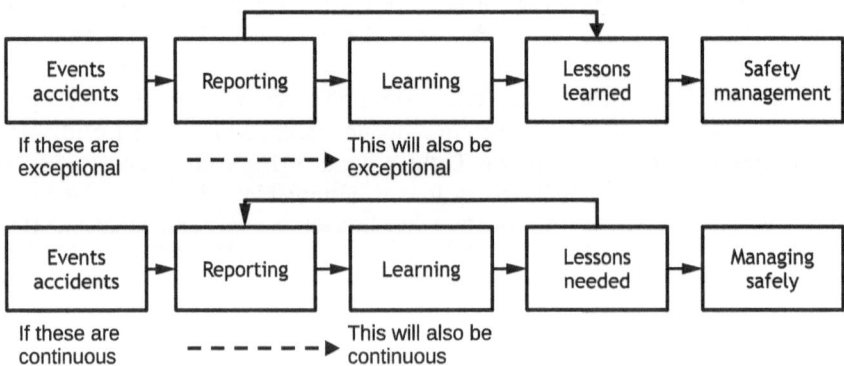

Figure III.2 Possible relations between reporting and learning

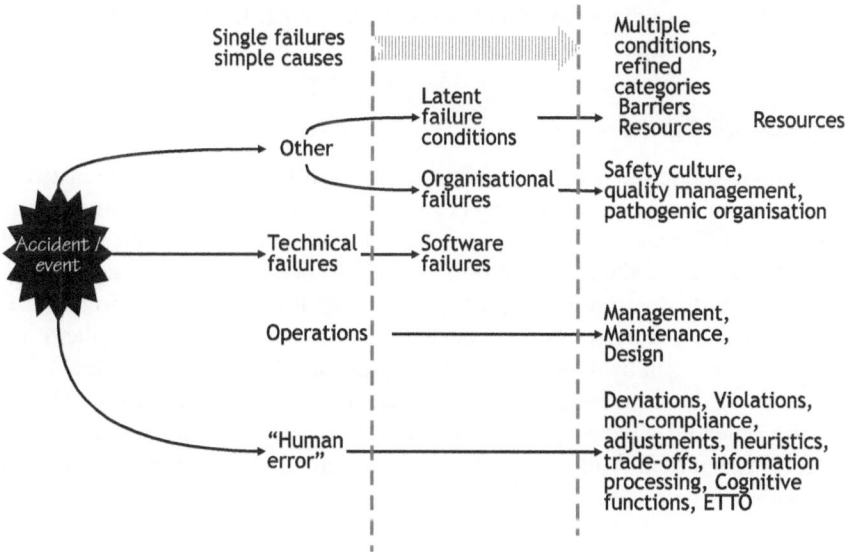

Figure III.3 Developments in types of causes

The categories for technical failures have seen less of a development, due in the main to more reliable systems. One notable exception is the category of metal fatigue, which became known to the general public when two British de Havilland Comet airliners crashed in 1954, two years after their introduction in commercial flight. The cause turned out to be the weakening of a metal part due to repeated cyclical movement such as bending or twisting. Since then the technology has improved to the extent that metal fatigue has disappeared from sight.

Another kind of technical cause, which seems to be more difficult to get rid of, is software failures or software errors. Although these usually are seen as a kind of technical failure, they might equally well be considered as a human or organisational failure. Leveson (1995) has provided an excellent treatment of software failures, and there is therefore no reason to go further into this topic here.

Finally, the category of "other" has undergone a remarkable development. During the 1980s and onward there has been a significant development in our understanding of causes that have to do with the working environment, especially the organisation. One development has been the introduction of the concept of latent conditions – or latent failure conditions – due not least to the work of Reason (1990b). Another has been the growing understanding of the importance of organisational factors, leading to the emphasis on safety culture and quality management. This field is still developing, and more will be said about that in Part V.

The large number of candidates for causes means that the root cause concept clearly is inadequate and also that the construction of strict causal models is nearly impossible. In the end, of course, we still need the category of "other" for the situations where we cannot really explain what has happened or cannot easily find a satisfactory cause.

A cynical definition of causes

The previous discussion has hopefully made it clear that it is no simple matter to define what a cause is. Philosophers have been working at that at least since the days of David Hume and have more recently been joined by practitioners, although with the pragmatic aim simply to be able to do something about accidents.

In line with the arguments contained in this book and the whole line of reasoning that permeates current behavioural science studies of accidents and failures, a cause can be defined as the identification, after the fact, of a limited set of aspects of the situation that are seen as the necessary and sufficient conditions for the observed effect(s) to have occurred. The cause, in other words, is constructed rather than found just as the label "human error" is a judgment made in hindsight (Woods et al., 1994, p. 210).

The nature of causes

Determining the cause of an accident is a psychological (social) rather than logical (rational) process. Causes and explanations are built rather than found, but causes are not like archaeological artefacts that exist buried underground and just wait for someone to stumble on them. A cause is nevertheless an artefact because it is a social construct, an idea that has been created and accepted by the people in a society and in more ways than one. First: the very idea or common conviction that causes exist is a widely shared social construct epitomised by the Heinrich dogma, second: the cause itself that is agreed upon is also by its very nature a social construct:

> the attribution of error after-the-fact is a process of social judgment rather than an objective conclusion.
>
> (Woods et al., 1994, p.4)

Or more explicitly:

> A "cause" is the determination after the fact, of a limited set of aspects of the situation that are seen as the necessary and sufficient conditions for the effect(s) to have occurred.
> A "cause" must have the following characteristics:
>
> • It can unequivocally be associated with a recognised system structure or function (people, components, procedures tasks, etc.).
> • It is possible and affordable to do something to reduce or eliminate it.

- It must conform to the current "norms" for explanations in relation to the corresponding age of safety (as described in Part II).
- The determination of the "cause" is a relative (pragmatic) rather than absolute (scientific) process. Causes in an important sense do not exist before we start to look for them. They are a good example of the confirmation bias and the What-You-Look-For-Is-What-You-Find principle.

(Lundberg et al., 2009)

attributions of error are a social and psychological judgment process that occurs as stakeholders struggle to come to grips with the consequences of failures. Many factors contribute to incidents and disasters. Processes of causal attribution influence which of these many factors we focus on and identify as causal. Causal attribution depends on who we are communicating to, on the assumed contrast cases or causal background for that exchange, on the purposes of the inquiry, and on knowledge of the outcome.

(Woods et al., 2019, p. 33)

One might also point out that determining the cause of what has happened is heavily influenced by the hindsight bias (Fischhoff, 1975), which demonstrates that the attribution of error after-the-fact is a process of social and psychological judgment rather than an objective conclusion

Social constructs

The previous quote from Woods et al. (1994) is another way of saying that a cause is a social construct (Searle, 1995) as described in Part I. This simply means that it is based on a collaborative consensus, a set of ideas shared by many people, rather than on observations and empirical evidence. In the case of safety these ideas are primarily the ideas of causality, as expressed by the Heinrich dogma, plus the widely shared position that safety is defined by the absence of accidents (Reason, 2000, p. 3). A cause represents a satisfactory explanation why an accident happened, and if enough people are satisfied it becomes *the* cause.

The illusion of analysis

The issues with causal analysis were clearly addressed already by Perrow (1984), even though the term *social construct* had not yet been introduced. Perrow had few illusions about the objectivity of accident investigation when he wrote:

Formal accident investigations usually start with an assumption that the operator must have failed, and if this attribution can be made, that is the end of serious inquiry. Finding that faulty designs were responsible would entail enormous shutdown and retrofitting costs; finding that management was responsible would threaten those in charge, but finding that operators were responsible preserves the system, with some soporific injunctions about better training.

(Perrow, 184, p. 146)

Based on this type of reasoning, which alas was not unusual then and is not today, it is understandable that estimates of "human error" as an attributed cause often passed 90 percent. But there was never an answer to who or what was responsible for the remaining 10 percent or who was responsible for all the cases when there were no accidents or incidents. Perrow had, not surprisingly, fully accepted a vision zero mindset, and he can hardly be blamed for that.

Investigating differently

The essence of investigating differently can be expressed simply in this way: Instead of investigating what happened to wit, the accident, to determine the possible causes in the specific case. It makes better sense and is more productive in the long run to look more closely at what *should* have happened than to investigate what did not happen, meaning what hindered the successful completion or execution of an activity in the specific situation, which conditions were not fulfilled, or which resources were not available – if you investigate accidents you can only learn what *not* to do or what should be prevented or avoided the next time the same activity is called for. But the efforts spent to do this are a cost, although this may well provide some satisfaction in the short run. Investigating what should have happened but did not happen can teach you what you *should do* and *what should be facilitated or changed* to ensure that the activity goes well the next time it is called for. This will on the one hand ensure safety – because an event cannot go well and fail at the same time – and on the other hand contribute to productivity and quality. The efforts spent to do this are therefore not a cost but an investment in better future performance.

How to investigate something that has not happened

It may well be asked whether it is possible to investigate something that has not happened and to learn from that. While this may seem logically impossible since what did not happen did *not* happen and there will therefore be no evidence or trace of it, it is by no means impossible to consider what *should* have happened. That is actually what we routinely do when we plan work and try to manage a work situation, for instance to determine the number of qualified people and other types of resources that are needed, etc. There is furthermore a practical method by means of which it can be done, namely the functional resonance analysis method (FRAM; Patriarca et al., 2017). Investigating what has not happened is finally no more mysterious than performing a pre-accident investigation, which is mostly a fancy term for risk assessment dressed up in the robes of the so-called new view of human error (Conklin, 2012).

Why learning is necessary

Learning is necessary for any organism or system to ensure the necessary flexibility or variability in responses. If the environment is static it is also perfectly predictable and stable; the same things will therefore happen again and again (like on Groundhog Day), and in that case it is both sensible and cost-effective to prepare answers or responses to every possible event and situation, and it is also practically possible because their number will be fixed and limited, and there is therefore no need to learn. These preparations may be made *intentionally* as part of system design (but only for design-based events, (DBE)), an example being the emergency operating procedures in nuclear power plants or other safety critical installations, or they may happen *evolutionarily* if the changes in the operating conditions are not too frequent or too fast. Learning is necessary because unless the responses can be adjusted or improved to match changing conditions there will sooner or later be situations (the dreaded beyond design base events, (BDBE)) where a response cannot be made, in which case control of a situation or development may be lost. Changes are necessary not only to the set of responses but also to the set of conditions, signals, indicators, or patterns that can be detected and recognised. This represents the potential to monitor. Anything that happens unexpectedly is a surprise, and even if a response in principle may exist – in the worst case it could be a generic response such as flee or withdraw – it may be impossible to carry it out in time, because of a lack of resources or materials or people, a recent example being the evacuations due to Greek forest fires in on the Island of Rhodes in the summer of 2023. It is in practice neither possible nor affordable constantly to be able to respond to everything without delay. Societies provide countless examples of that, like minimum waiting times for the fire fighters, an ambulance or treatment for a diagnosed condition or disease, such as cancer for instance or even a response from customer service, or the attendance of a clerk or a sales clerk person, to say nothing of a waiter in a restaurant.

Learning what not to do

Safety-I defines safety as a condition where as little as possible goes wrong or fails. Therefore, whenever something has failed we rush to the conclusion that we need to learn from it to make sure that it does not happen again, compare to Figure III.1. In contrast to that Safety-II defines safety as a condition where as much as possible goes well.

The Zero Accident Vision is a safety "philosophy" which states that nobody should be injured due to an accident. It is more an ideal and a way of thinking rather than a numerical goal; in terms of accident prevention strategies, the Zero

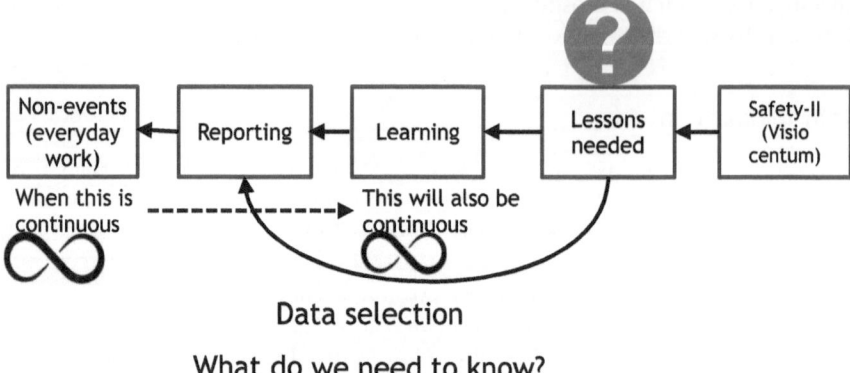

Figure III.4 Reporting as driven by learning needs

Accident Vision implies that all accidents can be prevented. When no accidents are allowed or approved, this in practice eliminates the basis for learning, hence invoking the regulator's paradox, but reporting could also be of that which went well rather than of that which failed; learning should in other words be based on all operations (Hollnagel et al., 2021) rather than on just failures and accidents. When reporting is exceptional, learning also becomes exceptional. The simple solution to dispense with reporting as a precondition to learning and instead engage in learning directly from all operations will quickly have significant consequences and also be less expensive; there will be more to analyse but it will also be easier to analyse.(Figure III.5). Today we mostly learn from what has gone wrong, because learning is triggered by reporting of what has gone wrong. This is described earlier as the Heinrich dogma. Yet reporting should be guided by the need to learn rather than the other way around. (See Figure III.5.)

Safety-I defines safety as a condition where as little as possible goes wrong or fails. Therefore, whenever something has failed we rush to the incorrect conclusion that we need to learn from it to make sure that it does not happen again, as compared to Figure III.1. In contrast to that, Safety-II defines safety as a condition where as much as possible goes well.

Learning what to do

Asking the question of *why* we want to learn rather than the other way around invokes the hypothesis of different causes. We conventionally learn from that which is the exception rather than the rule. Instead of having learning guided by reporting, where the reporting is usually limited to what has gone wrong or

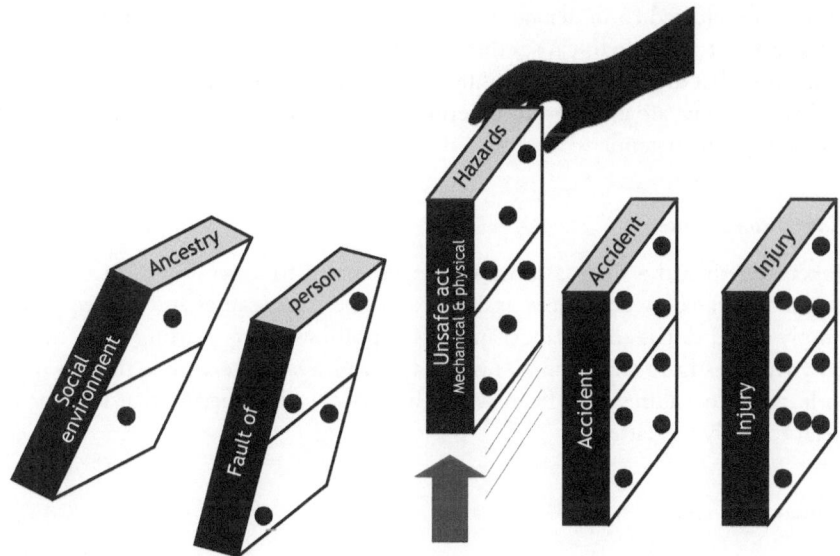

Figure III.5 The domino model based on Heinrich (1959, p. 16)

failed, reporting should be guided by what is needed to learn. We would therefore have to ask ourselves the question why we want to learn, as in Figure III.5. We think we should learn from accidents because we implicitly subscribe to the hypothesis of different causes.

When reporting is triggered by something exceptional happening, learning must wait for an accident to happen; this means that learning also becomes an exceptional activity. The basis for learning is also impoverished – the reporting acts as a filter for what actually happened in the situation, and the data analysis/investigation adds yet another filter to that. Learning is in this way limited to what the accident model defines as relevant and the recommendations are therefore determined more by the model than by the characteristics of the events being investigated.

The hypothesis of different causes

The safety legacy includes three important assumptions about symmetry between causes and consequences. The *first* is *the symmetry of magnitude,* which means that large consequences have correspondingly large causes and vice versa. In other words, large consequences cannot be due to trifles. It is this assumption that lies behind the notion of linearity. (But not the other form of linearity that is part of linear cause-effect reasoning in the Heinrich dogma.) The *second* is *the symmetry of valence,* which means that adverse consequences must be a result of errors or failures. The *third* is *the symmetry of complexity* – meaning that consequences that are uncomplicated to understand, such as an oil spill or a bridge that collapses, are assumed to be due to causes that also are simple to understand, whereas consequences that are complex and difficult to understand, such as an epidemic or global warming, are assumed to emerge from causes in ways that themselves are difficult to understand.

Causal consonance

Taken together, the three assumptions correspond to a hypothesis of similarity or consonance between causes and outcomes between causes and consequences, simply called causal consonance, and this hypothesis exerts a strong influence on how we investigate and learn and is also the reason why we assume there is more to learn from failures than from work that goes well, especially that there is no need to study the latter.

Causal dissonance

If we instead consider the converse of each of the three assumptions, that there are no symmetries of magnitude, of valence, or of complexity, that together corresponds to a hypothesis of different causes, meaning that the causes of unacceptable outcomes (work that fails) and acceptable outcomes (work that goes well) are different and that we therefore ought especially to pay attention to work that goes well as well as work that fails, corresponding to a principle of causal dissonance, which means that we must acknowledge the difference between resultant and emergent outcomes.

Asking the question of *why* we want to learn rather than the other way around invokes a hypothesis of different causes. We conventionally learn from that which is the exception rather than the rule. Instead of having learning guided by reporting, where the reporting is usually limited to what has gone wrong or failed, reporting should be guided by what is needed to learn; see Figure III.5. We must therefore ask ourselves the question of why we want to learn. Safety management and Safety-I agree with the principle of consonance between causes and consequences, in contrast to managing safely and Safety-II which agree with the principle of causal dissonance.

In this case the selection of data and information is based on the learning needs and the source is the near-infinite number (relatively speaking) of dynamic non-events where work goes well. In this way learning becomes the continuous activity it ought to be rather than the exceptional activity it in practice is now.

Learning what to do

The conclusion from safety centum is the exact opposite of the conclusion from vision zero, namely that when something has gone well we need to learn from it to make sure that it will also go well in the future (but we also need to learn from what has not gone well –indeed, we need to learn from all operations (Hollnagel et al., 2021)). The underlying argument is that in the macroscopic world an event or activity cannot go well and fail at the same time. (It can theoretically happen in a quantum world, but few industries operate there.) The outcome of an event may, of course, be judged from different perspectives, but they are usually not complete opposites. It is therefore essential to ask the question of *what* we need to learn – and *why*. Learning is one of the four systemic potentials (Hollnagel et al., 2021) that are described further in Part IV, and the essence of learning is improvement or change, to how a system responds, how it monitors, how it anticipates, and finally how it learns. Learning is necessary, since it is the only way in which it is possible to improve a system's responses.

The most efficient learner has the greatest chances of long-term growth and sustained existence. If a system always pays attention to the same signals, indicators, or trends and always responds in the same way to what happens, it will eventually perish. That may take place slowly and over a long time scale, as in the theory of evolution. But in today's turbulent society where changes to conditions, resources, and demands are rapid, learning must be equally rapid to be effective. Reporting may still be a prelude to learning, but reporting should be of that which goes well rather than of that which fails; learning should in other words learning be based on everyday operations. To overcome this problem it is necessary to engage in learning directly from all operations. This will both have significant consequences and be easier to do. We conventionally learn from that which is the exception rather than the rule.

There is a today a considerable number of accident models (Lehto & Salvendy, 1991; Wienen et al., 2017) and each model has an associated method which basically is a procedure for interpreting the facts of an event and mapping them onto the model.

Models and methods

Models are used everywhere in science but the term is often used in a very lose sense and used to glorify many arbitrary arrangements of boxes and arrows

> The basic defining characteristics of all models is the representation of some aspects of the world by a more abstract system. In applying a model, the investigator identifies objects and relations in the world with some elements and relations in the formal system.
>
> (Coombs et al., 1970, p. 2).

In consequence of this definition, whenever you are presented with something called a model, ponder whether it provides an answer to the following questions:

- Does it clearly identify objects or concepts that are essential to describe what is purported to be modelled?

- Does it make clear what the functional relations or dependencies are among these objects?

If you can answer neither of these questions, then it is not a proper systemic model in the scientific sense.

Monocausal accident models

The first model that is formally described is the domino model, which clearly is linear causality embodied. What came after are basically embellishments, although sometimes very elegant ones, but they never let go of the assumption that both causality and linearity are valid principles.

In the case of accident models, the aspects of the world are the causal factors or typical causes and how they are related to or affect each other. The clearest example of that is undoubtedly provided by the domino model (Figure III.5).

The visual rendering of the falling domino pieces is a compelling graphical argument that events unfold in a sequence represented by the relative position of the domino pieces. The physical analogy is immediately understandable. When one piece falls it will hit the next, which therefore also will fall. The physical analogy also presents the solution, to remove the third piece and thereby increase the distance between second and fourth pieces so the fourth piece is not hit by the second when that falls. Another solution according to the same analogy would be somehow to anchor the pieces to the surface on which they stand by a nail or by glue or replace them by dice, which are stable, but this is not as visually simple and striking as simply removing one piece. We euphemistically say that the model explains how the events unfolded and how the accidents happened. But that is not actually the case! Accident models do more than that; they also surreptitiously provide a ready-made solution based on the hidden assumptions of the models.

The domino model is an example of a monocausal description, meaning that a single cause, a domino piece falling, is sufficient for the effect to occur – the next piece falling. But it soon became obvious that real life was not as simple as that, meaning that a number of causes had to work together or coincide for the effect to happen. Most accident models today are therefore polycausal models.

Polycausal accident models

A polycausal accident model describes how a number of things (causes) have to coincide or combine in order to produce an effect. This, regrettably, clashes with the assumptions behind the three ages of safety thinking. A good example of a polycausal model is the bowtie model (Hale et al., 2004). Bowtie analysis is a simple diagrammatic way of describing and analysing the pathways of a risk from causes to consequences. Bowtie analysis (Figure III.6) is used to display a risk showing a range of possible causes and consequences. Although not as visually simple

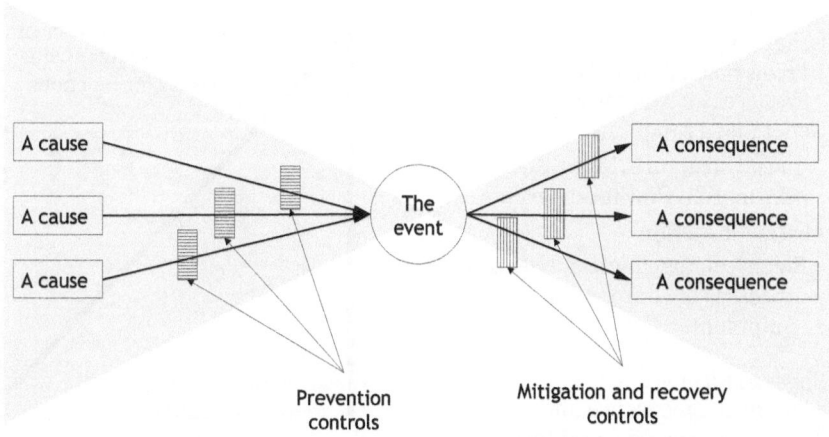

Figure III.6 Bowtie analysis from the ISO 31010 (p. 66)

or powerful as the domino model, Figure III.6 still shows not only that three causes have to coincide but also that multiple escalation controls or barriers must fail at the same time in order for the top event – or accident – to happen. The top event itself may also have multiple consequences which become realised if the mitigation and recovery controls fail. The best known example of a polycausal model is probably the Swiss cheese model (Reason et al., 2006). But there is a sequence implied even in a polycausal model. And it still leaves unanswered why the initial harmful event or cause happened. The domino model obviously has the same problem and unless this can be answered it is incorrect to say that the accident has been analysed. The problem was addressed by Heinrich but not actually solved by calling the first domino falling the root of the trouble as Heinrich (1931, p. 106) did, even though that is no more an explanation than the inflation field was in the big bang theory. In the Swiss cheese model, the first slice "fallible decisions" remains unaccounted for; perhaps fallible decisions just happen randomly? The answer to the infinite regress of the first or root cause, that even Woods and Cook (2002) could not solve, is possibly that it really is "a case of turtles all the way down."

Layered accident models

In the case of polycausal accident models the question soon arrives of whether and how the multiple causes are related, specifically how many layers there are and how they relate to each other since it clearly is naive to assume they are mutually independent. In the Swiss cheese model for instance there are five layers (or slices), called fallible decisions, line management, psychological precursors of unsafe acts or latent failures, unsafe acts, and finally inadequate defences. One of the more popular derivatives of the Swiss cheese model, the human factors analysis and classification system (HFACS) framework (Shappell &

Functional purpose
  Production flow models
  System objectives, constraints, etc.
Generalised functions
  Causal structure. Mass, energy, and
  information flow topology, etc.
Abstract function
  Electrical, mechanical, chemical
  processes of components and
  equipment
Physical form
  Physical appearance and
  anatomy; material and form;
  locations, etc.

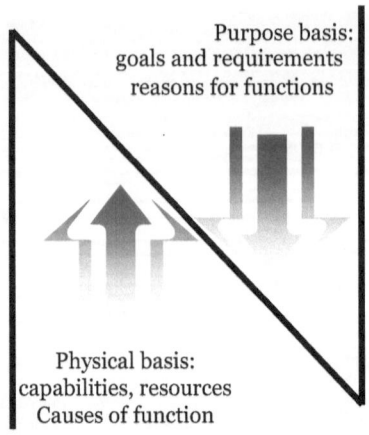

Purpose basis:
goals and requirements
reasons for functions

Physical basis:
capabilities, resources
Causes of function

Figure III.7 The abstraction hierarchy from Rasmussen and Lind (1981)

Wiegmann, 2000), has only four layers or slices, called organisational influences, unsafe supervision, preconditions, and unsafe acts respectively, ending with an accident, but with holes in all four layers begging the question of what a hole or deficiency of unsafe supervision possibly could be, it looks suspiciously like just an illogical double negative.

The holes in the original Swiss cheese model are in the final layer only; the other four layers represent a sequence of events or failures, not too unlike Heinrich's domino pieces. The question is, of course, what comes before the first slice or first domino piece. Heinrich foresaw this problem, and his proposed solution was to name the five domino pieces of the model (although he called them accident factors rather than domino pieces):

> The first domino piece was called *social environment*. (Could this be a precursor to safety culture?)
> The second domino piece was called *fault of person* (but not human error since that concept was not in use at the time).
> The third domino piece was called *unsafe act*.
> The fourth domino piece was called *accident*.
> The fifth and final domino piece was called *injury* – the consequences of the accident.

The distinction between the second and the third piece is interesting, not least in comparison with contemporary terminology (Table III.2)

Both accident models fail to account for the occurrence of the first factor (domino piece) or the first slice of cheese.

Table III.2 Semantics of the domino model

Name of domino piece	Heinrich (1931) examples	Swiss cheese equivalent	Contemporary equivalent
(1) Ancestry and social environment	Environment may develop undesirable traits of character or may interfere with education. Both inheritance and environment cause faults of person.	Psychological precursors of unsafe acts or latent failures	Organisational (safety) culture
(2) Fault of person	Inherited or acquired faults of person; such as recklessness, violent temper, nervousness, excitability, inconsiderateness, ignorance of safe practice, etc., constitute proximate reasons for committing unsafe acts or for the existence of mechanical or physical hazards.	Unsafe acts	Human error
(3) Unsafe act and/or mechanical or physical hazard		Inadequate defences	Failure mode or phenotype
(4) Accident	Events such as falls of persons, striking of persons by flying objects, etc. are typical accidents that cause injury.	Accident	Accident, top event
(5) Injury	Fractures, lacerations, etc. are injuries that result directly from accidents.	Harm	Consequences

The best known layered model is unquestionably the AcciMap which is based on Jens Rasmussen's abstraction hierarchy (Figure III.8) first described by (Rasmussen & Lind, 1981) and later by Rasmussen (1987). In the original publication it was described as follows:

> In the abstraction hierarchy, the system's functional properties are represented by concepts which belong to several levels of abstraction, see fig. 4 [in the original publication]. The lowest level of abstraction represents only the system's physical form, its material configuration. The next higher level represents the physical processes or functions of the various components and systems in a language related to their specific electrical, chemical or mechanical properties. Above this, the functional properties are represented in more general concepts without reference to the physical process or equipment by which the functions are implemented, and so forth. At the lower levels, elements in the process description match the component configuration of the physical implementation.
>
> (Rasmussen & Lind, 1981, p. 12; according to this description Heinrich's domino model might arguably be seen as a rudimentary version of an abstraction hierarchy.)

However in a later critical review of the abstraction hierarchy, the second author wrote the following:

> It is concluded that the semantics of the means–end levels and their relations are vaguely defined and therefore should be improved by making more precise distinctions. Furthermore, the commitment to a fixed number of levels of means–end abstractions should be abandoned and more attention given to the problem of level identification in the model-building process. It is also pointed out that attempts to clarify the semantics of the abstraction hierarchy will invariably reduce the range of work domains where it can be applied.
>
> (Lind, 2003, p. 67)

Epidemiological accident models

Polycausal and multilayered accident models are sometimes also called epidemiological models (another term is a H(ost)-A(gent)-E(nvironment) HFACS model). The allusion is the analogy to an epidemic. In this analogy an infection/injury/damage is due to the interaction among host, agent, and environment. In an epidemic (such as COVID-19) the host organism becomes infected when exposed to a carrier or an infectious agent, a virus that its defences or barriers (think immune system) are ineffective against an. The Swiss cheese model is clearly an epidemiological model.

An epidemiological model is thus radically different from a monocausal accident model, approaching but not quite achieving what an emergent accident model could accomplish.

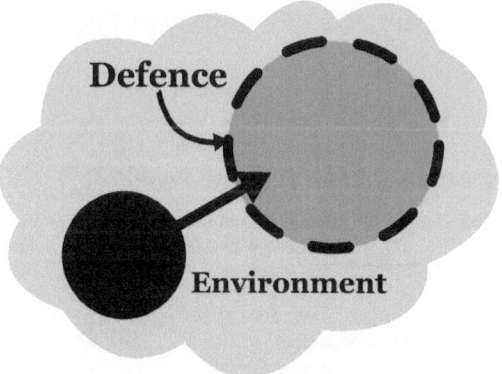

Figure III.8 The H(ost)-A(gent)-E(nvironment) model accident model

Model–cum–method

In practice most accident models represent a model–cum–method unit, in the sense that the method comes with the model and in practice is built in even though that is rarely recognised or acknowledged by the model (developers). We have, however, become so used to this lack of precision that we fail to notice that an investigation method in fact is defined by the model even when it pretends to be just a method. The ideal would be a method that was just a method, for instance a method without a built-in model or a method-*sine*-model; without that an accident investigation cannot really be useful. At the time of writing the only such method seems to be the functional resonance analysis method (FRAM; Patriarca et al., 2017).

Accident investigation is futile, because causes do not exist before we look for them – or rather before we impose them on the facts in accordance with our preferred accident model. Causes thus represent a *world-as-imagined*, not a *world-as-it-is* or even a *world-as-it-was*, when the event took place. The lessons learned and the conclusions drawn from such investigations will therefore have limited practical value when applied to actual situations of work and in that way defy the very purpose of investigations. In consequence of the Heinrich dogma, and vision zero, and the widely accepted truth that we must learn from accidents (Kletz, 2001), reporting has effectively become a precondition to learning (Figure III.9).

In this case reporting is triggered by something exceptional (an accident or incident) happening; learning must wait for an accident to happen, and since accidents mostly are rare and exceptional, learning also becomes rare and exceptional, contrary to what it ought to be. The basis for learning is also impoverished – the reporting acts as a filter and only provides a simplified description of what actually happened in the situation, and the data analysis/ investigation is yet another filter on top of that. In this way learning becomes limited to what the accident model defines as relevant and important, hence

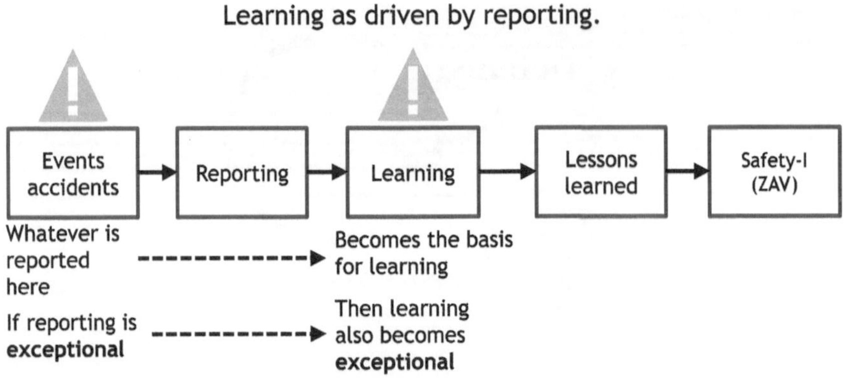

Learning as driven by reporting.

Events accidents	→	Reporting	→	Learning	→	Lessons learned	→	Safety-I (ZAV)

Whatever is reported here - - - - - - - - - → Becomes the basis for learning

If reporting is **exceptional** - - - - - - - - - → Then learning also becomes **exceptional**

Figure III.9 Reporting as a precondition to learning

merely reinforcing the model's built-in assumptions. But *investigating differently* may have value and yield practicable results, in the sense that it can tell us what *to do* and what to support and facilitate rather than what *not* to do and what to prevent or eliminate

This limitation to learning can fortunately easily be overcome by removing reporting as a prelude to learning – which thereby becomes more than avoidance learning – and instead engaging in learning directly from all operations. Doing so will have significant consequences. We conventionally learn from that which is the exception rather than the rule. Instead of having learning guided by reporting, where the safety legacy limits reporting is to what has gone wrong or failed, reporting should be guided by what is needed to learn (Figure III.4).

In this case the selection of data and information is based on a recognition of *what* it is necessary to learn and *why*. The source is furthermore the (near) infinite number of cases where work goes well. This eliminates the need to wait for something to happen before learning can begin. In this way learning becomes a continuous activity and an integral part of everyday operations as it ought to be.

Futility epilogue

The overall conclusion of Part III is that the purpose of an investigation is to build a cause rather than to find it. Causes, of course, exist even before an investigation is made. But unlike physical artefacts they do not passively lie hidden waiting to be found. Because causes are social constructs they can only be found in one place, namely in the collective mindset of the people who carry out the investigations, where they probably have been buried for years and might better have been left undisturbed.

It is an illusion coming from the safety legacy that an accident has a root cause and it is a delusion that we can find it, in many cases the cause/causes, hence the recommendations from an investigation, are determined more by the assumptions built into the accident model than by what actually happened.

By investigating accidents you only learn what you should *not* do or what you should avoid. But in the long run it is not only more valuable but also easier and more motivating to learn what you *should do* and become better at. To achieve that it is necessary to pay attention to (investigate gives the wrong connotations) work that goes well. And there is fortunately plenty of it.

IV Systemic potentials management

Management prologue

Management is a synthesis of the functions needed to prepare, organise, and control the resources of a system or a company so that it can perform as required. The latter is important because the goals for management – and the criteria for acceptable performance – are usually external to the management process itself. Management can in principle serve two different purposes:

- The first purpose is to *maintain* a current state or position by compensating for potentially destabilising internal or external influences.
- The second is to *attain* a new position either by approaching a more desirable position or state or by distancing oneself from a current undesirable position or state.

In order to do so, management logically requires three types of knowledge (in a parallel to the reasoning of Mackay (1956):

- First, knowledge about the *goals*.
- Second, knowledge about the current state or *position* (where you are now such as the current state or condition of the system or company).
- Third, knowledge about effective ways or *means* to change position relative to the goal, knowing how you can move from where you are now to *where* you want to be in the future, including *when* you want to arrive.

Managing something is very often by means of a travel or voyage metaphor. It is thus common to talk about how to keep or improve a position or how to get closer to or reach a target and even to envisage a road map for the way ahead. A voyage metaphor is convenient since it clearly is essential to be able to control how something moves and changes position, whether the movement is actual or metaphorical (such as the safety culture voyage, Tiessen, 2008) and whether that which moves is tangible (as an automobile, a vessel, or an aeroplane) or intangible (as safety or quality). (The literal meaning of tangible is that something is perceptible by touch.) The metaphor is also useful because it points to the need for the three different types

DOI: 10.4324/9781032664729-4

of knowledge listed earlier and described further later. To illustrate the concepts, imagine you are driving in another country (in Europe) or in another state (in the US). In this case it is clearly important to know both your current position and your destination. And if you are driving a car, steering a ship, or flying a plane, knowing the means (how to change your position) is not an issue. The metaphor is certainly relevant for tangible or first-order systems since it here makes sense to describe movements in a physical or material sense. But it is less appropriate for intangible or second-order systems even though it remains widely used. The voyage metaphor is unfortunately seriously misleading in the sense that it tempts us to think about the path into the future as a map, such as a road map.

Road maps versus sea charts

The limitation of a road map is that it shows a static landscape, road works, blizzards, earthquakes, and wars excepted, where the road ahead is clearly indicated, both on the map and in the physical world, whereas the travel in the case of change management is more like a voyage at sea where there are no fixed indication of where to sail except in archipelagos and narrow straits with heavy traffic so it is constantly necessary to determine your position and plot your course, neither do the charts have any indication of whether the sea ahead will be calm or rough or whether there are waves wind, icebergs, floating containers, or other hindrances that may create difficulties for the intended voyage.

First-order and second-order performance

In relation to management it is useful to make a distinction between the first-order functions or performances of a system or a company, which are the raison d'être

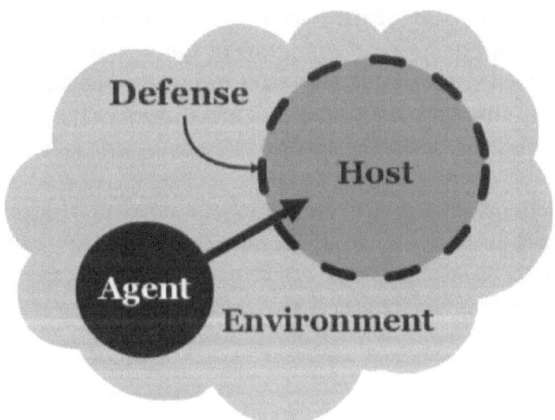

Figure IV.1 The H(ost)-A(gent)-E(nvironment) model accident model

of the system or company, how it generates the revenue needed to keep going, and the second-order performance, which is the way the system or company manages the first-order performance. The first-order performance is in the hands of the people at the celebrated sharp-end, while the second-order functions or performances are in the hands of the people at the equally celebrated blunt end.

At the sharp end people manage their own work, and through that also the safety of what they do at the local workplace, traditionally referred to as the front line or coalface (whether it is safety in the sense of Safety-I or Safety-II). Here the processes being managed are tangible because they are easy to understand and concrete and they are also reasonably fast so single-loop learning (which means making direct adjustments to a process or development to correct a mistake or a problem) will suffice (single-loop learning is basically the classical feedback loop that governs practically all human behaviour). The work processes, whatever they are, usually have a physical presence and are also reasonably fast, and what happens is in most cases subject to – or constrained by – known principles or such as the laws of physics. Appropriate strong signals can therefore be defined, although they may need to be supplemented by other, weak signals. But exactly the opposite is the case for those who work at the blunt end who manage the organisation as such, hence the work of others rather than their own work. For them safety rarely is an issue at the local workplace. If an accident happens, they will not be harmed beyond possibly losing their bonus or, in the worst case, their job. Managers need to ensure both that the first-order activity (whatever the organisation does to stay in business) is acceptable and that the second-order activity (how the organisation itself performs) also goes as required. In the latter case the process being managed is intangible because it is vague or abstract, and changes usually happen so slowly that single-loop learning is ruled out. Management must therefore rely on double-loop learning for which there are few if any concrete correlates or concomitants, it is difficult or even impossible to define or understand, and it may even be complex (cf., Part II). There is therefore limited scope for or benefit of single-loop learning. Neither is there much basis for defining strong signals, so weak signals must suffice. The bottom line is that everyone, regardless of what they do and what their role in a system or a company is, must rely on a combination of strong and weak signals, simply because it is impossible completely to specify everywhere what the relevant strong signals are. The weak signals are the bits of information and performance patterns that by experience are known to be useful in order to perform with the requisite agility. For management single-loop learning cannot be used because changes happen so slowly that any feedback will be too late for practical use (Annett, 1972); management must instead depend on double-loop learning where the adjustments are of the goals or decision-making rules that govern single-loop learning, hence a feedback loop around another feedback loop (Argyris, 1977; Figure IV.2). Double-loop learning will in practice always be defeated by delays in the feedback from organisational developments which may easily be an order of magnitude slower than the work processes that are the focus of single-loop learning; this feedback is not

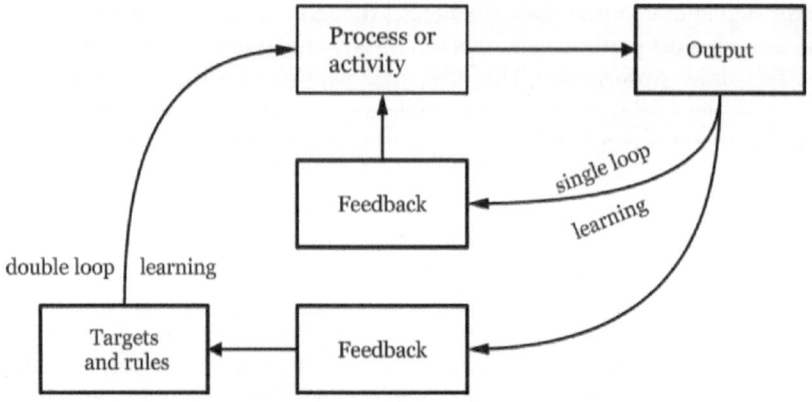

Figure IV.2 Single and double-loop learning based on Argyris (1977)

actually usable for management processes and organisational changes. (There is an interesting parallel between double-loop learning and von Förster's second-order cybernetics, mentioned in Part II).

Since the purpose of management is to ensure an adequate basis for work at the sharp end, wherever that may be, it is clearly just as important to understand the work-as-done of managers as it is to understand the work-as-done of sharp-end workers and operators. And just as the weak signals that sharp-end workers use have been studied, so the weak signals that managers use should also be studied. The practical question is what the important weak signals for management are and how these signals can be made easier to recognise. Strong signals are the distinctive and disruptive events – usually in the form of reported accidents and incidents at the sharp end – that should rather not have happened. Management blunders rarely have immediate effects and therefore cannot serve as strong signals. Strong signals at the sharp end are well defined, unmistakable, attract attention, and are therefore difficult to miss. But they represent the absence rather than the presence of safety (Reason, 2000, p. 3) and are therefore neither suited for the purposes of safety management, nor the purposes of managing safely.

Signal detection theory

Signal detection theory (Peterson et al., 1954) is concerned with the ability to differentiate between information-bearing patterns (called signals) and random patterns (called noise) that distract from and sometimes hide the signals. This is usually treated as a question of the detection threshold of the signal, hence the S/N ratio but in a psychological rather than a physical sense. Here countless studies have shown that human operators are not passive receivers of information but rather active decision-makers and information seekers who ceaselessly make challenging perceptual judgments under conditions of uncertainty. Weak signals are therefore more about the meaningfulness of signals than their strength.

First type of knowledge: what is the position?

Before beginning to make a change, it is obviously necessary to know the initial position, regardless of whether the change is a movement in physical space or in a metaphorical (safety, quality) space. In addition to knowing the initial position, it is also necessary to know how the position changes and therefore to know the current position at any particular time. Only by comparing the position at different times is it possible to determine whether the change (or movement) goes in the right direction and at the right rate.

Strong and weak signals

Strong signals represent the information that a priori has been defined as necessary for effective safety management. Strong signals are what people know or have been told they must pay attention to and therefore also what they notice – as described by the what-you-look-for-is-what-you-find (WYLFIWYF) principle (Lundberg et al., 2009).

Weak signals, on the other hand, are ambiguous and may be difficult to detect because they have limited predictability, fail to attract attention, or are seen as insignificant both physically and psychologically. The weak signals comprise the many small events that lie below the threshold of seriousness and therefore are not reported but also include the usually unacknowledged performance patterns – the habits, the routines, and the common trade-offs – that most of the time lead to the expected outcomes, A weak signal has been defined as:

> A seemingly random or disconnected piece of information that at first appears to be background noise but can be recognized as part of a significant pattern by viewing it through a different frame or connecting it with other pieces of information.
>
> Schoemaker and Day (2009, p. 86)

Weak signals as sources of information are not considered by conventional safety management systems and are not seen as relevant performance indicators, let alone key performance indicators (KPIs). Weak signals comprise what we are not prepared for, what we do not expect, and what we fail to notice for other reasons. The strong signals represent the information, the KPIs that are acknowledged as necessary; either based on a theory or based on experience. Yet a weak signal may every now and then in hindsight be linked to unexpected and unwanted results. Weak signals therefore in many ways correspond to the patterns of "dynamic non-events" that Weick (1987) argued were the foundation of reliable performance. Weick also noted that such performance is invisible in the sense that reliable outcomes are constant (and therefore below the jnd, described next), which means there is nothing to pay attention to.

In a different field, the interest for weak signals arose in the mid-1970s in connection with the growing interest in "strategic management" and "strategic surprises." Ansoff (1975) put forward a conceptual framework and a practical procedure that would allow a firm operating in a turbulent environment to plan for strategic surprises by responding to weak signals, while (Pastore & Scheirer, 1974) in a more conventional application of signal detection theory looked at how it could be applied to the study of cognitive processes.

A weak signal is something that is easily missed because it is not recognised as meaningful in the current context, hence a psychological rather than a physical phenomenon. The ability to recognise patterns whether as signs, symbols, or signals is crucial for the proper functioning of joint cognitive systems in any dynamic setting (Rasmussen, 1983). Patterns, however, cannot be idiosyncratic but must represent a social consensus within the given frame of reference. Patterns can point to solutions – as in recognition-primed decision making (Klein, 1998) – yet still allow problems to be solved individually by fitting them to the current conditions and context (Woods et al., 2021). System design and system management must therefore pay attention to how people make use of patterns or weak signals at work.

Accidents and failures are the traditional strong signals both by their meaning and by their visibility. Experienced operators at either the sharp or the blunt end have through experience found out what the important weak signals are, which is one more reason it is essential to pay attention to what actually happens when "nothing" happens and work goes well. Weak signals are, however, not just the bits of information that do not fit the apparent patterns and configurations of the available information but also the observable *regularities or patterns* in how work is done. These regularities in turn arise from habits, routines, roles, rituals, social conventions, or physical constraints and may often not even be recognised by the performers themselves. Weak signals are important not only for front-line operators, but they also play a role in what people do everywhere. Cases where the outcome is an unexpected useful discovery are happily accepted as the result of serendipity (Merton & Barber, 2011). But in most other cases the use of weak signals is simply called recognition, as in the ways people rely on their experience both to make sense of the situations they are in and to decide what to do (Klein, 1993).

Weak signals, jnd, and change blindness

In the 1860s the psychological study of perception – or psychophysics – introduced a concept called the just noticeable difference (jnd; in psychology it is described by the eponymous Weber-Fechner Laws, the phenomenon is also known as the difference limen, the difference threshold, or the least perceptible difference). The jnd (or the Weber fraction) defines how much a stimulus must change in order for a difference to be noticeable in at least 50 percent of the cases. If a change is smaller than the jnd, it is likely to go unnoticed, hence constituting a psychologically weak signal. As a simple example, consider the two alarm panels shown in Figure IV.4. In Panel A there is just a single alarm, and it is therefore easy to notice when it lights up. But if the same alarm tile lights up in Panel B, it will be less easy to notice because there already are 20 other active alarms. What is a strong signal in condition A becomes a weak signal in condition B. This applies to visual as well as to auditory alarms, which have no spatial position. The jnd provides an analogy to weak signals in change management (Ansoff, 1975). The jnd can, however, not be used literally because system performance is about discrete outcomes and qualitative changes, where psychophysics is about quantitative – and mostly continuous – changes of input stimuli. In change management, the issue is not directly the sensitivity of some sensory organ, unless management metaphorically is described as a sensing organism. Change management is rather about the ability to notice whether a change of some kind has taken place. Going beyond psychophysics, people may sometimes fail to notice what happens even when a distinct change is introduced in a visual field or stimulus if it coincides with some other change (a condition called change blindness (Simons & Rensink, 2005)). Since this is a dominant feature of visual perception, it can partly explain the weak signals problem at both the sharp and the blunt ends. But beyond the deficiencies in noticing changes in the information that is presented and stable in a situation, there is also a critical issue in noticing changes or events that take place over time.

Spatial and temporal patterns

When weak signals are defined as random bits of information, the allusion is that some of these can be missed because of the failure to notice a change or because of the inability to recognise an overall pattern. The weak signal is in this case something that does not immediately correspond to a Gestalt or a schema – or possibly worse, something that is mistakenly perceived as being present because it matches an assumed Gestalt or schema (as a kind of perceptual or cognitive confirmation bias). A Gestalt is a unified whole with properties that cannot be derived by summation from the parts and their relations usually as a geometrical organisation or configuration of individual items (of information) coming from various sources or placed at different spatial locations. A schema (Wagoner, 2013) is a pattern of thinking or comprehension that is used to interpret what happens in and around the system world; the concept was introduced by Sir Frederic Bartlett in the 1930s and is today used by computer science and artificial intelligence. It is also an important concept in cognitive psychology (Neisser, 1967). An accident model is a good example of a schema. Gestalt psychology has formulated a number of principles or laws – proximity, similarity, figure-ground, continuity, closure, and connection but only for visual Gestalts – that describe

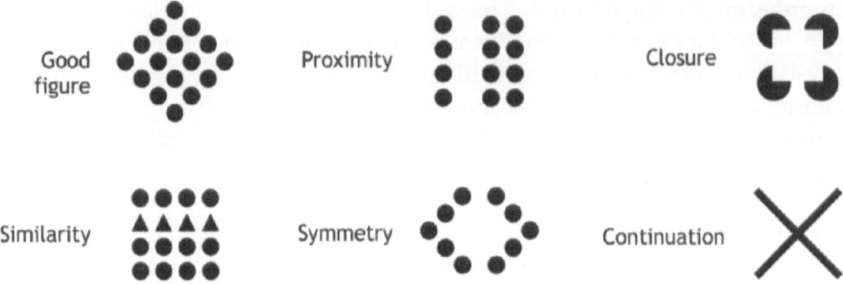

Figure IV.3 Illustration of six classical visual Gestalt principles

how the human brain perceives and organises visual elements. Six well-known visual Gestalt principles can be seen in Figure IV.3.

Temporal patterns as weak signals

But a pattern or Gestalt may also be a relative temporal position (the order in which events happen, such as the order in which different events happen or the timing of events, so it is a question about different temporal positions rather than how they are organised spatially. The set of the logically possible temporal relations (but not quite temporal Gestalts, because they are cognitive rather than perceptual; they are not immediately recognisable but require some kind of reasoning) is shown in Table IV.1 and in Figure IV.4. Several of these – guess which? – unfortunately tempt people to reason *post hoc ergo propter hoc* and in that way incorrectly to infer a causal relation where there is none, even though it by now is widely known that a correlation does not imply causality.

As long as things happen reasonably fast, relative to the human ability to remember and keep things in mind, temporal and spatial patterns can be considered as analogous. But if things happen very slowly, temporal patterns gradually dissolve. This is easy to illustrate by music. If a piece of music (such as Für Elise) is played very slowly (for instance *larghissimo* instead of the *poco moto* Beethoven prescribed, and it is no problem to find concrete examples on the web), it can be difficult to recognise the melody even when you know what it is. An extreme example of a stretched temporal pattern is a composition by the American composer and philosopher John Cage aptly named "As Slow As Possible." Here patterns are clearly recognisable in the written music, even for people who cannot read music. But when played very slowly, the patterns vanish because we are unable to hear adjacent sounds "together" – almost as a kind of temporal change deafness. (The composition is currently performed at an organ in the St. Burchardi Church in Halberstadt, Germany, in strict compliance with its title. The performance commenced on September 5, 2001, with a pause lasting until February 5, 2003. The first chord was then played until July 5, 2005. The planned duration of the performance is no less than 639 years! It cannot practically be much

Table IV.1 Temporal relations (Allen, 1983)

Temporal relation	Meaning (example)
Before	The first event takes place before the second event and they do not overlap.
After	The first event is after the second event and they do not overlap. This is equivalent to before with the events reversed.
During	The first event happens during the second event so that the first event starts after the second event has started and ends before the second event ends.
Meets	The first event begins before the second event and there is no interval between them, i.e., the second event begins when the first event ends.
Overlaps	The first event begins before the second event ends and continues after the second event has ended.
Equal	The first event and the second event start and end at the same time.
Starts	The first event begins at the same time as the second event but ends before the second event does.
Finishes	The first event starts after the second event has started but ends at the same time as the second event.

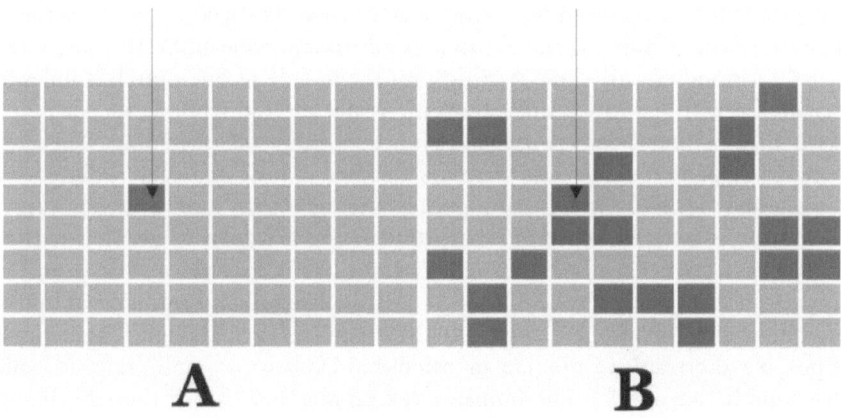

A   B

Figure IV.4 Noticeable (A) and unnoticeable (B) differences

slower. Needless to say it is not performed by a human organist, neither is there a permanent audience, although enthusiasts drop by from time to time to listen for a brief moment to this unique performance.) Another example of temporal patterns is provided by time-lapse photography. This is a technique where the frequency at which film frames are captured is much lower than the frequency that is used to view the sequence. When shown at normal speed, time appears to

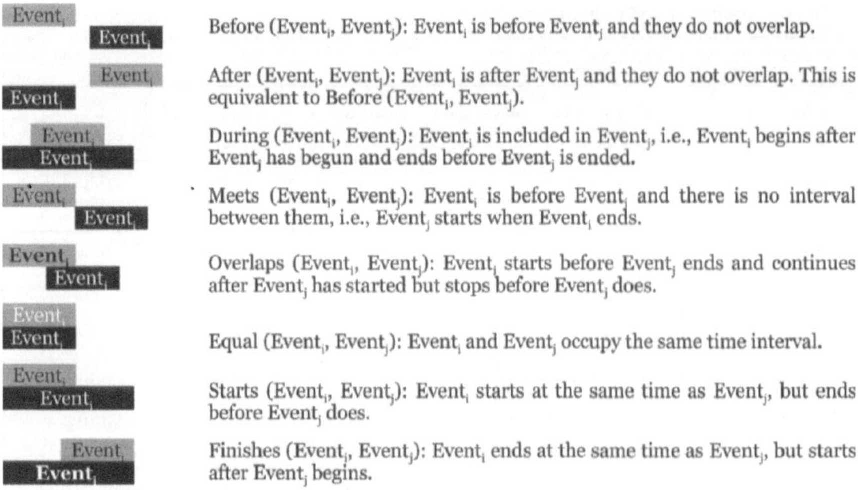

Event$_i$ / Event$_j$	Before (Event$_i$, Event$_j$): Event$_i$ is before Event$_j$ and they do not overlap.
Event$_i$ / Event$_j$	After (Event$_i$, Event$_j$): Event$_i$ is after Event$_j$ and they do not overlap. This is equivalent to Before (Event$_j$, Event$_i$).
Event$_i$ / Event$_j$	During (Event$_i$, Event$_j$): Event$_i$ is included in Event$_j$, i.e., Event$_i$ begins after Event$_j$ has begun and ends before Event$_j$ is ended.
Event$_i$ / Event$_j$	Meets (Event$_i$, Event$_j$): Event$_i$ is before Event$_j$ and there is no interval between them, i.e., Event$_j$ starts when Event$_i$ ends.
Event$_i$ / Event$_j$	Overlaps (Event$_i$, Event$_j$): Event$_i$ starts before Event$_j$ ends and continues after Event$_j$ has started but stops before Event$_j$ does.
Event$_i$ / Event$_j$	Equal (Event$_i$, Event$_j$): Event$_i$ and Event$_j$ occupy the same time interval.
Event$_i$ / Event$_j$	Starts (Event$_i$, Event$_j$): Event$_i$ starts at the same time as Event$_j$, but ends before Event$_j$ does.
Event$_i$ / **Event$_i$**	Finishes (Event$_i$, Event$_j$): Event$_i$ ends at the same time as Event$_j$, but starts after Event$_j$ begins.

Figure IV.5 The logically possible temporal relations between two events (from Allen 1983)

tun faster so that very slow changes, such as the blossoming of a flower – or even of paint drying – become visible. More generally, two events can only be experienced together if they both exist within the span of working memory, in the sense that the "first" event is still remembered when the "second" event happens. The exception to this is if the events are really memorable and if the purported causal relation between them is critical for the people in question. Examples of that are easily found in weather observations and age-old wisdom, such as "Red sky at night, sailors delight. Red sky in morning, sailors take warning." This presumably helped sailors to anticipate the weather, at least in a very rudimentary sense, and it also illustrates the strong human tendency to "find" or invent causes and relations to make sense of what happens and not least to be prepared for what may happen. Farmers have similar sayings, like "Clear Moon, frost soon." In the lack of strong empirical evidence or an articulated theory, this kind of thinking belongs to folklore and approaches superstition, even though it sometimes is possible afterwards to provide an articulated explanation (and preferably one that sounds "scientific"). The human working memory can be thought of as a mental time window, where the leading edge is the "now" and the trailing edge is the "then" and where it only is possible to perceive or connect what is inside the frame (between the two edges the "then" and the "now." The precise nature of our working memory and the actual capacity is no simple matter Cowan (2008). For the present discussion we simply acknowledge that there is a limit to how much a person can consider "together" and propose that the working memory span is in the order of two to three hours; it is hopefully considerably longer for organisations – even though no one really knows or seems to be concerned about it. The working memory capacity is fixed for humans, but for organisations it may be possible artificially to extend it. Events that are separated by a span exceeding the working memory capacity because they happen more slowly do in some sense not "exist" together in working memory and it will therefore be impossible to notice any patterns (momentous events excepted, as mentioned earlier).

This can be illustrated as shown in Figure IV.6. Here the first-order activities take place on a time-scale measured in minutes or hours and refer to what sharp-end operators must pay attention to and manage. This will typically be tangible processes where events occur so frequently that it is easy to recognise patterns.

Second-order activities, such as organisational changes are at least an order of magnitude slower and take place over weeks or months, sometimes even years so that earlier events will typically be forgotten when a new event occurs, as illustrated by Figure IV.7. Here the frame that represents working memory span is too narrow to contain more than one event at the same time. Because of that, it is impossible for anyone to perceive even two events "together," unless something is done artificially to facilitate that. The temporal patterns do logically exist, but they are outside the range that we can easily be recognised. Temporal patterns are at least as important as spatial patterns when it comes to understanding the weak signals that people rely on in their work. Noticing the order or sequence in which things happen is the basis for inferring patterns (primarily hypothetical cause-effect relations) in what happens. We assume that people act with purpose and intention and that this together with the instructions or procedures determines what they do and how they do it. In relation to individual

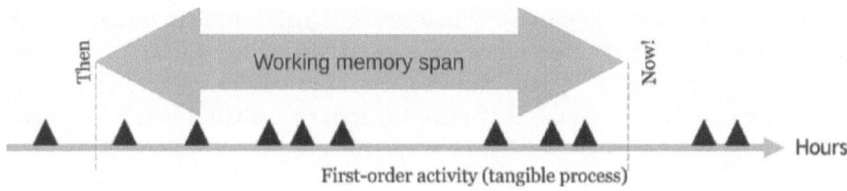

Figure IV.6 Working memory span compatibility with first-order activities

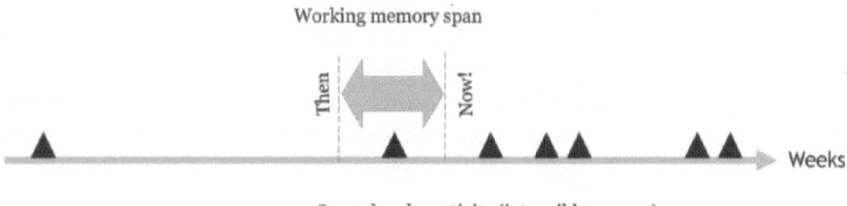

Figure IV.7 Working memory span compatibility with second-order activities

performance, this becomes a person's modus operandi – the characteristic patterns of work-as-done – that others may be able to recognise, make sense of, as a weak signal. In relation to collective performance this becomes the typical way in which a system or a company works, both the routines of daily activities and how it responds to the unexpected (finding patterns in safety-related events developments and situations). Temporal patterns are therefore essential sources of information about performance, for individuals as well as organisations. For the management of the first-order or tangible processes – such as controlling the movement of a vehicle (an aircraft, a ship, or a car) – the speed by which things happen, and therefore also the extent of the temporal patterns, is determined by the nature (and dynamics) of these processes. Since people only can manage processes for which they themselves can recognise temporal patterns, it follows that their own performance, how they respond, in turn will exhibit corresponding temporal patterns that others may then perceive. The situation is, however, completely different for the management of the second-order or intangible processes that provide the basis for safety, quality, or reliability. For tangible, first-order processes, we know which information and which changes are important – such as the movements and positions of vehicles – while for intangible second-order processes we do not know or seem to care much. The underlying processes are not only vaguely defined, as illustrated by the many problems in finding meaningful strong (safety) signals or KPIs, they are also slow in the sense that developments and changes take a long time – which means that there are no easily recognisable temporal patterns as discussed earlier. (It may be likened to listening to "As Slow As Possible.") Yet since experience clearly demonstrates that safety management cannot rely exclusively on the traditional strong signals such as accidents and reportable events, because these are strong signals for the absence of safety, with no corresponding strong signals for the presence of safety, (cf. Reason (2000), p. 3) it becomes necessary to define some form of appropriate weak signals. Moreover, since the weak signals cannot be inferred empirically from observable performance, they must be derived analytically from principles and concepts such as resilience. This will be discussed in the following.

Second type of knowledge: what is the goal?

In order to know whether a change goes in the right direction and at the appropriate rate, it is necessary to know what the goal or target is and when it should be reached. This is also necessary to determine when and whether the change or movement has been completed. The goal should therefore be described in practical or operational terms, preferably absolute and concrete rather than relative (thus not as better than "X" or better than last year). While this is straightforward in the case of tangible systems and material processes (such as the movement of aircraft or the production of goods) it is less easy in the case of intangible systems and more abstract movements, such as a higher level of safety or an improved safety/learning/reporting culture, etc.

Third type of knowledge: what are the means?

The third type of knowledge is about effective means – how concretely to make a change, how to change the current position either to get closer to the desired position or goal or to get away from an unwanted or unacceptable state or position, for instance with too many accidents. In the case of tangible systems – guiding vessels, manufacturing some product, transmission of information, energy, or materials – we usually know how to make the change because the process takes place in or by a physical system that has been designed and provided with the necessary means of measurement and control. But that is not the case for changes that refer to intangible systems, to concepts, or to abstractions. (The controls that can be used to manage an organisation are indirect, hypothetical, and often disputed.) Different counties and different leaders may for instance have completely different ideas of how to control inflation; most will rely on raising interest rates but one, President Erdogan of Turkey, thought otherwise and favoured inaction, with disastrous results. What means are available for changing safety? For improving quality? For increasing precision or reducing delays? For improving the culture? Few effective means are known, and there is little agreement or evidence that the often favoured approach of increased regulation and standardisation is actually the best, thus leaving the field open to anyone with sufficient self-confidence. Despite many claims to the contrary, organisations are neither designed nor built but grow almost organically from an initial idea, shaped by forces of which we know little, despite decades of research and several impressive theories. If we knew what actually happened, it might be possible to develop effective means beyond trying to change the hearts and minds of people (Hudson et al., 2002).

Making changes: the basic feedback loop

It is not difficult to find advice on or recommendations for how to make a change. This is probably because making changes to how something works, is controlled, or is organised has been necessary since the very first societies and organisations, such as armies, large commercial activities (trading houses), public administration, collection of taxes, and health care. The management literature is replete with recommendations which when looked at more closely all turn out to be variations on the same basic theme, namely the feedback loop seen in Figure IV.8.

The essence of feedback, as the name implies, is that information about the output or outputs or results of a process or activity is used to control or adjust the process or activity that produces them. Feeding back simply means giving something back to where it came from. The term *feedback* is of modern origin and became commonly used in the 1950s. But the principle that feedback describes is much older. The first known example of a mechanical device using feedback was a float valve that would regulate the inflow of water to fill a tank, so that the flow automatically stopped when the tank was full, thereby ensuring a constant level of

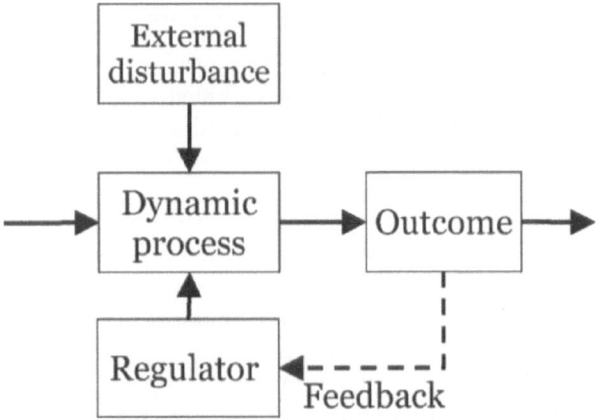

Figure IV.8 The basic feedback loop

water; it is basically the same device we have in our toilets today to stop the inflow when the reservoir is filled. It was constructed by a Greek inventor and mathematician called Ctesibius who lived about 200 BCE in Alexandria in Ptolemaic Egypt, but it was at that time used to regulate the outflow for a water clock. Ctesibius is also recognised as the inventor of the pipe organ.

Feedback is an essential part of the common scientific paradigm or the *hypo-thetico-deductive* method. According to this, scientific inquiry proceeds by formulating a hypothesis in a form that can be falsified, using a test on observable data where the outcome, the feedback, is not known in advance. A hypothesis is an articulated assumption or prediction about something, which can be verified and perhaps also falsified by comparing it to empirical evidence. The basic scientific method is usually attributed to the British philosopher Francis Bacon, and it appears in print in 1620. The relevance of this for change management is that a hypothesis essentially is a prediction of certain causal relationships, hence what the results will be if this or that is done – or not done, as the case may be.

The essence of the scientific method is as follows: A test outcome that runs contrary to predictions of the hypothesis is taken as a falsification of the hypothesis. A test outcome that could have but does not run contrary to the hypothesis verifies the theory (although one prominent philosopher of science (Popper, 1959) has argued against that). It is then proposed to compare the explanatory value of competing hypotheses by testing how stringently they are corroborated by their predictions using Occam's razor. The hypothesis thus has the same function or role as the plan in most change paradigms.

Feedback and feedforward

Feedback means that information about the consequence of actions and interventions derived from systematic observations or measurements is brought – or fed – back to the process that is being managed so that corrections and adjustments can be made on the basis of actual experience. But it is, of course also possible to make corrections and adjustments based on expected change, hence the presumed or predicted rather than the actual results. This is unsurprisingly called feedforward. The advantage of feedforward is that the need to wait for results, hence any response delay disappears. Feedforward is anticipatory or proactive rather than reactive control. The benefit can be a significant gain in time, which in competitive cases provides the ability to surprise opponents (in war) and the competition (in business). The disadvantage is that actions will be based on predictions and therefore critically depend on the correctness of the predictions or more precisely on the correctness of the assumptions, model, or theory from which the predictions are derived. Feedback is certain because it refers to something that is known and that has happened. Feedforward is uncertain because it refers to something that is unknown because it is yet to happen.

The danger of being caught up in a hypothetical world and ignoring the feedback was vividly described in Stefan Zweig's novella *Chess Story* (*Schacknovelle*).

Feedforward serves as a prescription or plan for a feedback, which the actual feedback may or may not confirm. A plan, as in PDSA cycle, which is described later, is a good example of feedforward, but the plan and the model behind it must be confirmed by feedback. Feedforward represents the theory underlying a hypothesis; feedback represents the evidence.

Change management could in principle overcome the delay problem for double-loop learning by being based on feedforward rather than feedback. But this would require far better articulated management theories than even Stafford Beer's viable systems model (Beer, 1984).

Test operate test exit (TOTE)

The classical feedback loop is conceptually simple because it does not describe the process or activity that produces the output. It had a renaissance in the mid-1950s and was given another name, TOTE, in a book with the interesting title *Plans and the structure of behaviour* (Miller et al., 1960). The main argument in this book was that behaviour was not just a set of involuntary responses to stimuli as behaviourism and reflexology maintained, but that something happened in between, namely the formulation of a plan and an intention for what to do; the same idea appeared many years later in the so-called activity theory (Engeström et al., 1999). The TOTE was illustrated by a very simple activity, namely hammering a nail into a piece of wood. Here the process (operate) is the hammering with the hammer – and hitting the nail, rather than the finger! – until the head of the nail is flush with the piece of wood (test). Once this has been achieved, the process can end (Exit), as shown in Figure IV.9.

In terms of the scientific model mentioned earlier, the hypothesis here is simply that when you hit the nail with the hammer it will go further into the piece of wood, and the likelihood of this hypothesis being true is very high. It can also

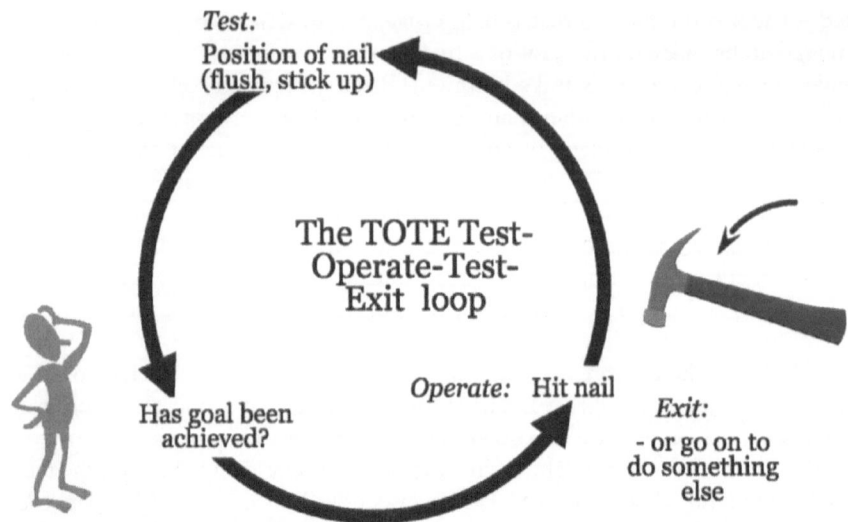

Figure IV.9 The TOTE loop (from Miller et al., 1960)

easily be verified by most people. The plan is also fairly simple and will bring about the intended outcome – unless you hit the nail askew so that it bends.

Although organisational changes are in practice never as simple as hammering a nail into a piece of wood, the same principles apply: There must be a plan for what to do (to reach a goal) and a hypothesis about how it can be done (the *operate*) and also a way to assess the current position during the process (the *test*).

Making changes: PDSA

Possibly the best known paradigm for making changes is the Plan–Do–Study–Act or PDSA cycle, usually attributed to Deming (1950), who derived it from the Plan–Do–Check–Act (PDCA) cycle by Shewhart (1931) including obligatory nods to both Galileo Galilei and Francis Bacon (Moen & Norman, 2010). By a strange coincidence Shewhart's book on quality control was published the same year as Heinrich's book on industrial accidents. But this is probably no more than that – a strange coincidence.

The basic PDSA cycle looks like this, Figure IV.10. The following description of the PDSA is taken from the website of the W. Edwards Deming Institute:

> The cycle begins with the Plan step. This involves identifying a goal or purpose, formulating a theory, defining success metrics and putting a plan into action. These activities are followed by the Do step, in which the components of the plan are implemented, such as manufacturing something. Next comes the Study step, where outcomes are monitored to test the validity of

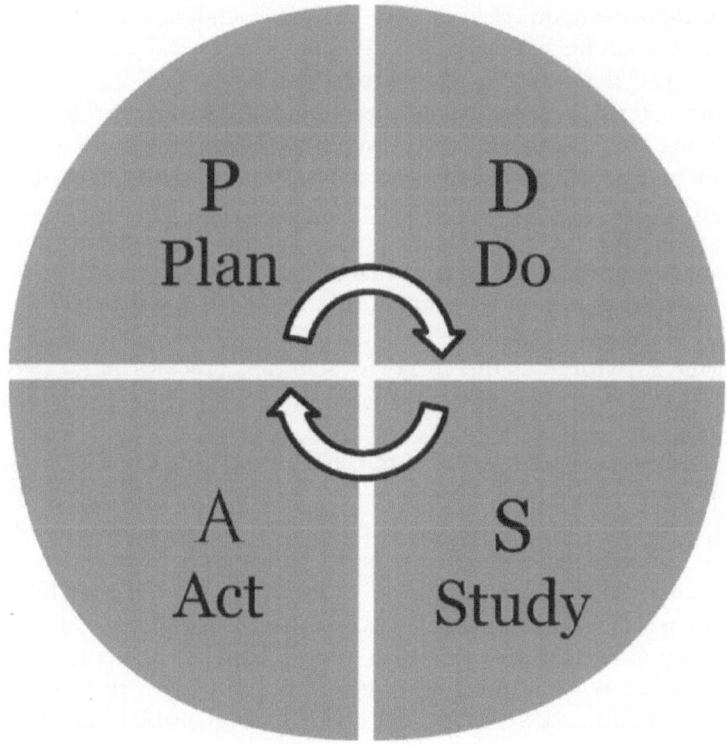

Figure IV.10 The PDSA cycle from Deming (1950)

the plan for signs of progress and success, or problems and areas for improvement. The Act step closes the cycle, integrating the learning generated by
the entire process, which can be used to adjust the goal, to change methods
or even reformulate a theory altogether. These four steps are repeated over
and over as part of a never-ending cycle of continual improvement.

The similarity to the hypothetico-deductive method is striking and by no
means coincidental

- **Plan**: Establish the *objectives* and *processes* necessary to deliver results in
  accordance with the expected output (the *target* or *goals*). By establishing
  output expectations, the completeness and accuracy of the *specifications* is
  also a part of the targeted improvement. When possible start on a small
  scale to test possible effects. In the TOTE this is the intention to drive a
  nail into a piece of wood, probably to serve some other purpose, like
  hanging a picture or something else. (In which case the nail certainly
  should not end being flush with the wood.)

- **Do**: Implement the *plan*, execute the process, make the *product*. Collect *data* for charting and analysis in the following "check" and "act" steps. In the TOTE this is hitting the nail with the hammer.
- **Study**: Study the *actual results* (measured and collected in "do" step) and compare them against the expected results (*targets* or *goals* from the "plan" step) to ascertain any differences. Look for *deviation* in implementation from the plan and also look for the appropriateness and *completeness* of the plan to enable the execution, i.e., "do." Charting data can make this much easier to see trends over several PDCA cycles and in order to convert the collected data into information, that is what you need for the next step "act." (In Shewhart's original version of the cycle, "study" was called "check." Walter Shewhart was the creator of statistical process control (SPC)). In the TOTE this is checking whether the nail has gone sufficiently far into the wood.
- **Act**: If the check shows that the plan that was implemented in the "do" step resulted in an *improvement* to the prior standard (*baseline*), then that becomes the *new standard* (baseline) for how the organisation should act going forward (new standards are enacted). If the check shows that the plan that was implemented in "do" is *not an improvement*, then the already existing standard (*baseline*) will remain in place. In either case, if the check showed something different than expected (whether better or worse), then there is some more learning to be done, and that will suggest potential future PDSA cycles. But it would generally be counter to PDSA thinking to propose and decide upon alternative changes without using a proper plan phase or to make them the new standard (baseline) without going through do and check steps. In the TOTE this step does not exist; it is what the person does after the hammering is finished.

Critical issues of the PDSA

Critical issues relating to "plan": How thorough should the planning be? How long can it take? How many details and hypothetical scenarios should be considered (given the ceteris paribus limitations)? What is the time horizon; when must the planning be finished? The meaning of ceteris paribus is that all other things are equal. It is an assumption traditionally made in controlled experiments and hypothesis testing, but it is rarely fulfilled.

Critical issues relating to "do": How long will it take to implement something before the results become manifest? Does it require a simple and single action (such as hitting a nail)/activity or a composite one (such as changing the hearts and minds of people)? Does it include a single process or multiple processes? If multiple, are they independent or interdependent? And are they synchronous?

Critical issues relating to "study": Can the actual outcomes or results be reliably counted or measured? Are the outcomes immediate or delayed? Have clear indicators (KPIs) been defined? Is there anything to compare them to? And are there clear evaluation criteria?

Critical issues relating to "act": How can the improvements be implemented and made permanent? This is clearly not an issue in the case of hammering a nail into a piece of wood. (Kurt Lewin addressed it by the change meta-model of "unfreezing, change, and refreezing" described in the following section on action research.) Will the new standard be appreciated and adopted by people at the sharp end? How can you ensure that everyone involved knows what they are supposed to do differently (this may be tricky in the case of cultural changes)? Will everyone be willing to make the change and discard old habits? How do you make sure that the necessary resources are available? Will the benefits from the change outweigh the costs?

Historically, the PDCA/PDSA were intended for use with quality control of industrial manufacturing, where most of these questions could easily be answered, since machines so far rarely have attitudes or develop habits. But today the PDSA is used mostly for changes in socio-technical systems, although that was never intended. It would be a miracle if it actually worked. Yet miracles do not happen very often in our century. Many of the questions are neither meaningful nor can they be answered for socio-technical systems, because humans both individually and collectively are radically different from machines.

The list of critical issues could go on. Another important part of controlling a change is knowing how long it will take, or at least to have a reliable estimate of the duration. This, of course, requires a detailed understanding of the dynamics of what actually goes on in an organisation. Knowledge about how much time a change will require is essential both when detailed plans are made, when means of intervention are chosen, and when resources are assigned. It is also necessary to be realistic about possible side-effects (cf. Merton's law of unanticipated consequences described in Part II) and about unplanned outcomes, in particular if these can be detrimental. The better the system or process is known and understood, the fewer unanticipated side effects there will be and vice versa. There is fortunately a solid basis for governing changes, known as the Law of Requisite Variety, described in a following section.

Making changes; Observe – Orient – Decide – Act (OODA)

Since the iterative process of putting forward a hypothesis and then testing it is enshrined in the fundamental scientific paradigm, it is hardly surprising that there are several other versions, such as Observe Orient Decide Act (known as the OODA loop), developed by John Boyd, a colonel in the US Air force.

The OODA Loop is (again) a four-step process for making effective decisions in high-stake situations such as aerial combat with an antagonistic adversary. It involves collecting relevant information (observe), recognising potential biases (orient), deciding, acting, and then repeating the process with new information. The basic OODA

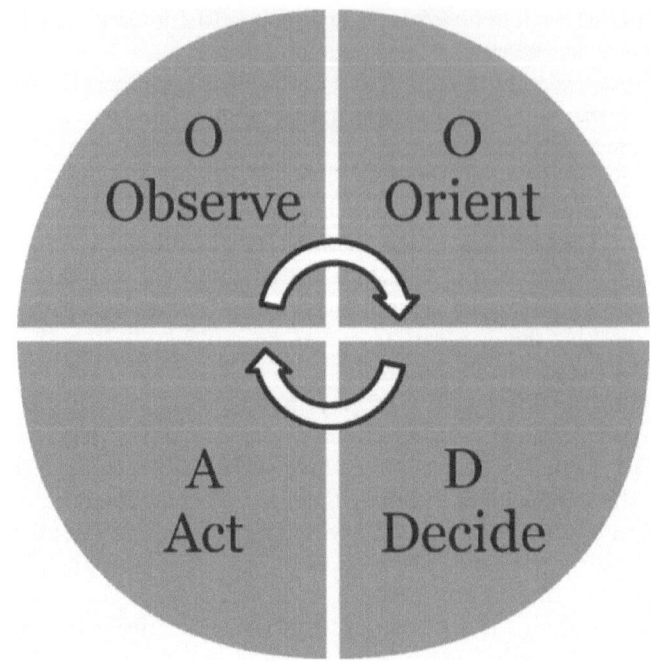

Figure IV.11 The OODA loop

loop can be seen in Figure IV.11. It basically has the same parts as the PDSA. *Decide* matches "do," *observe* and *orient* match "study," and *act* finally matches "act."

But the OODA loop was intended for rapidly changing conditions such as aerial combat with an active adversary unlike the PDSA which assumed basically stable, hence fully predictable, environments such as production lines. The OODA therefore made a virtue of not sticking to the same underlying model or understanding of the situation but revising it from time to time.

"To comprehend and cope with our environment we develop mental patterns or concepts of meaning . . . we destroy and create these patterns to permit us to both shape and be shaped by a changing environment" (Boyd, 1976, p. 1). This advice might actually not be amiss for more peaceful forms of change management.

And Boyd continued:

> The uncertainty and disorder generated by an inward-oriented system talking to itself can be offset by going outside and creating a new system . . . uncertainty and related disorder can be diminished by . . . creating a higher and broader more general concept to represent reality.
>
> (Boyd, 1976, p. 6)

As if to emphasise Boyd's arguments, Brehmer (2007) further developed the OODA loop to become a dynamic OODA (or DOODA) loop.

Action research as change management (the spiral of steps)

Since the cycle in both PDSA and OODA has to be carried out repeatedly, the approach is in both cases a set of cycles, and since each cycle is intended to produce some progress or improvement it is obvious to think of it in terms of a spiral or, even better, a helix. And that is indeed the intention of yet another method, developed by the German psychologist Kurt Lewin, probably best known as the father of action research. The method is called spiral of steps, where each step is a cycle of planning, action, and fact-finding about the result of the actions shown in Figure IV.12.

Lewin, however added something important to the other cycles of change. He worked in Berlin in the 1930s and might therefore not have known about the PDCA or the PDCA. Lewin published his spiral of steps idea in 1946. He had fled the Third Reich and lived in the US in the 1940s working at Cornell university, so he may well have been aware of Shewhart's book by then. Kurt Lewin, however, developed his ideas in a completely different context, such as action research illustrated by his classic three-stage change meta-model of "unfreezing, change, and refreezing" (Lewin, 1952), which is unfortunately is missing from the other change paradigms. The purpose of the "unfreezing" stage is to create an awareness that the status quo in some way limits the organisation and that a change is therefore needed. (This is critical for social systems but not relevant for the manufacturing processes that Shewhart and Deming had in mind). The "changing" stage is the transition from one way of working to another and may involve the assimilation of new technologies. It is during the changing stage that people hopefully adopt new behaviours, adjust to new conditions, and get used to new ways of thinking. The third and final stage is the "refreezing" – or just "freezing" – where the changes gradually become institutionalised and therefore become the new "normal." The purpose of the "refreezing" is to prevent people from reverting to old habits once the transition phase is over since that would

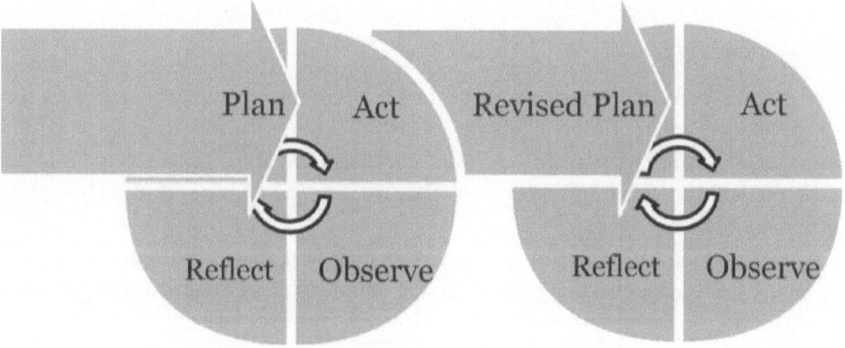

Figure IV.12 Lewin's action research cycles (from Santally et al., 2015, p. 5)

quash the effects of the change. Lewin studied physics before he did psychology, and the analogy is said to be based on the fact that an ice block cannot be forced into a new shape without breaking. Instead, to achieve a transformation from one shape to another, it must first be melted (unfrozen or thawed), poured into a new mould (changed), and then frozen again in the new shape (refrozen).

The law of requisite variety (LoRV)

In order to manage an organisation and indeed in order to manage or control anything, it is necessary to understand how it works — to understand why things happen the way they do, particularly as a result of deliberate changes and interventions. Apart from being good common sense, the need to understand how something works is also "one of the really fundamental laws of cybernetics" (Beer, 1966, p. 279). It was formulated in cybernetics in the 1940s and 1950s (Ashby, 1956), where it is known as the Law of Requisite Variety (LoRV). The law simply states that the variety of the outcomes (of a system or a process) only can be decreased by increasing the variety in the controller of that system. Another way of expressing that is the Good Regulator Theorem, which states that "every good regulator of a system must be a model of that system" (in practice this means that the regulator must be able to refer to a model of the system; Conant & Ashby, 1970). A more formal rendering (see Figure IV.13) is that Min $(V_O) = V_D - V_R$ where V_O is the variety of the output of a system, V_D is the variety of the system, and V_R is the variety of the regulator. In plain language the LoRV simply says that "a control system, or regulator, must have as many possible states as the system it is to control – it must be able to recognise and respond to any condition or situation that may occur." In other words, the variety of the outcomes (of a system) can only be decreased by increasing the variety of the regulator of that system. Effective control is consequently impossible if the regulator has less variety than the system or process (see the regulator's paradox in Part I) it is supposed to control.

Types or levels of control

Quite apart from cybernetics and the Law of Requisite Variety, it stands to reason that one can be more or less in control of a situation or development, that there can be different types or levels of control. One attempt of describing that is the proposal to distinguish between four different modes of control, Table IV.2 (Hollnagel, 2002).

In other words, if something happens in a system that either cannot be recognised by the system management or for which management cannot provide a response, then control will inevitably be lost. Modern societies can unfortunately provide nearly countless examples of that. The essence of this law is illustrated in Figure IV.13, where the icons are used to illustrate the (imperfect) match

Table IV.2 Control mode characteristics

Control mode	Number of goals	Choice of next action	Evaluation of outcome
Strategic	Several	Prediction based	Elaborate
Tactical (attended)	Several (limited)	Plan based	Normal details
Tactical (unattended)	Several (limited)	Association based	Perfunctory
Opportunistic	One or two(competing)	Association based	Concrete
Scrambled	One	Random	Rudimentary

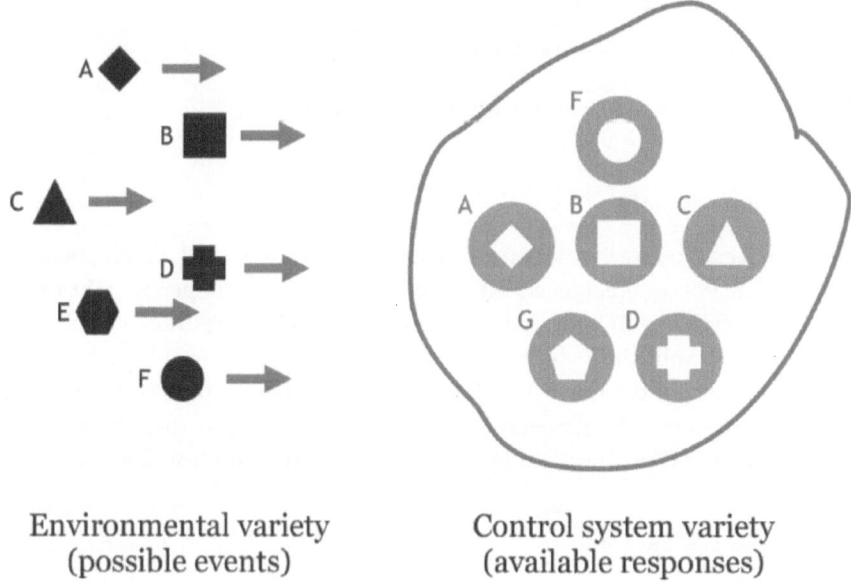

Environmental variety   **Control system variety**
(possible events)   **(available responses)**

Figure IV.13 The Law of Requisite Variety (illustration)

between environmental variety and control system variety. In this example there is no response if E happens. Here the event E cannot be recognised by the regulator, and there is no response available if and when it happens (it may not even be detected!).

The LoRV is unfortunately often misinterpreted in the following sense, as in Figure IV.14. To many people the simplest solution to reduce V_O is to reduce the variety of the process V_D instead of increasing the variety of the regulator V_R – but this will not work in the long run, because such reductions may themselves introduce additional variety, unless the process is perfectly known in every detail.

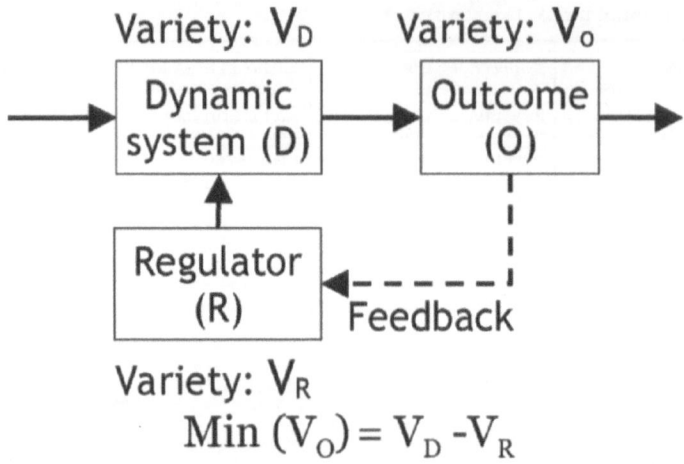

Figure IV.14 The Law of Requisite Variety (principles)

The Bacon cycle?

The similarity between the various models for change and development is striking but not a coincidence. They were all developed by people (Shewhart, Deming, Lewin, Boyd) with a scientific education and background who all must have known about the hypothetico-deductive method. (This is not meant as undue criticism of the four but rather a suggestion that it might be appropriate to honour the one who started this four centuries ago.) So, perhaps they should all be called the Bacon cycle to honour Francis Bacon who first described it?

How can changes be made?

The inevitable question is whether there is any other way of making a change to a complex socio-technical system. In an engineering system, a machine, you can make a change by replacing a single component with one that is improved in some way, but socio-technical systems and the societies we create are not machines, however much we like to think so. Machines have been designed and built so we know them to the tiniest detail and understand how the parts or components interact; in Perrow's terminology they are intentionally loosely rather than tightly coupled and simple rather than complex. (Otherwise we would not have been able even to think of them, let alone design them!) Unlike machines, organisations and socio-technical systems are not built but rather grow following principles we hardly understand, so we simply do not know enough about them to change them in any other way, even though we desperately wish we could and sometimes pretend we can. (As countless cases demonstrate, you cannot improve how a department or a company performs simply by replacing the current leader with a better one.)

The substitution myth

Socio-technical systems differ from the manufacturing facilities Shewhart and Deming developed their methods for because they do not obey the principle of interchangeability or the *substitution myth* (Bradshaw et al., 2013), according to which a part or component of a system can be replaced by an identical part without adverse effects. The depressing conclusion of this is that we probably cannot do much better than we do now. The hypothetico-deductive method has after all proved its worth since it was first described in print in 1620, and there ought to be some merit to that.

A minor difference between the hypothetico-deductive loop and the PDSA cycle is that the former does not include an act step. Yet scientific developments usually lead to some action, in most cases is a publication of some kind, today more than ever so, as a thesis, a journal paper, a conference presentation, a podcast, and in extreme cases even as a book.

The four systemic potentials

First systemic potential: respond

Understanding *how* something happens is necessary in order to know what to do and when to do it – which means knowing how to respond. It is evidently impossible to control something unless an appropriate response can be made to whatever happens. This either requires that a response already exists (together with the necessary resources) – or that one can be developed before it is too late. The inability to respond, whether it is two moving vehicles, or business initiatives on a potential collision course, or a rapidly spreading bushfire, or a pandemic (like COVID-19), will eventually lead to a loss of control. Quite apart from that, no one likes to find themselves in a situation where they cannot respond to what happens.

It is therefore an advantage if the number of unexpected events is as small as possible and if as much as possible goes well. If something is expected to happen – even though the exact timing may be uncertain – then it is possible to prepare in advance or even to act pre-emptively, although that (as a form of feedforward) may be a risk in itself. In order to be prepared for what may happen – at the next moment or in the immediate future – it is necessary to keep an eye on what happens in the system as well as around it.

Second systemic potential: monitor

Understanding how something happens is necessary in order to know what to look for (signals and trends), where to look, and how often – which means knowing how to monitor. Although it sometimes may be possible to respond to unexpected events, "fire-fighting" is not a strategy that is recommended in the long run.

> In business organisations, there are invariably more problems than people have the time to deal with. At best, this leads to situations where minor problems are ignored. At worst, chronic fire-fighting consumes an operation's resources. Managers and engineers rush from task to task, not completing one before another interrupts them. Serious problem-solving efforts degenerate into quick-and-dirty patching. Productivity suffers.
>
> Bohn (2000)

Under such conditions the number of unanticipated consequences is likely to increase.

Third systemic potential: learn

Understanding how something happens is also necessary in order to know what the relevant experiences are and where they can be found – which means knowing what to learn (already mentioned in Part III). It is through learning that a system (or a company) is able to change how it responds to something, by reinforcing effective responses, by learning new responses, or by suppressing ineffective ones. The essence of learning is not the acquisition or accumulation of knowledge (or data or information) as such, but the effect this has on the ability to respond – as well as the ability to monitor and recursively also the ability to learn. Without the ability to learn, responses would be limited to a fixed and predefined set. Yet, always responding in the same way is only feasible if conditions never change, i.e., if the world is fixed and perfectly stable, even if you nominally are too big to fail. But as we all know the world is not perfectly stable. The same argument, of course, goes for monitoring. It is unwise always to rely on the same indicators and measurements. Learning can furthermore not be limited to what *not* to do (avoidance learning). It is equally important, if not more so, to learn from what works well, in order to reinforce effective responses. Learning is necessary to make performance more efficient, as in the progression from knowledge via rules to skills (Vicente & Rasmussen, 1988). The gradual automation of responses; the development of patterns, habits, and routines; and the recognition of weak signals are essential parts of how performance becomes attuned to the prevailing conditions – with all the risks that such reduced thoroughness and increased effectiveness involve.

Fourth systemic potential: anticipate

Understanding how something happens is finally necessary in order to know what to take into consideration and be concerned about before it happens – which means knowing how to anticipate. Where monitoring is concerned with keeping track of what happens here and now and what may happen in the immediate future based on current trends and recognised temporal patterns, anticipation is concerned with that which has not yet happened but which may. One form of anticipation deals with the intended (and anticipated) consequences of decisions made and of actions taken, to say nothing of the unanticipated outcomes. When something is done to achieve a specific outcome, anticipation serves as a "what-if" game or scenario: If we do so and so, then X, Y, or Z will happen – and also what will happen if we do nothing depending, of course, on the conditions and on what others may do.

De Laplace's demon

Anticipation would be much easier if we could assume that strict determinism existed as believed by the French mathematician Pierre Simon, Marquis de Laplace. In 1820 de Laplace wrote that:

> We may regard the present state of the universe as the effect of its past and the cause of its future. An intellect (a demon) which at a certain moment would know all forces that set nature in motion, and all positions of all items of which nature is composed, if this intellect were also vast enough to submit these data to analysis, it would embrace in a single formula the movements of the greatest bodies of the universe and those of the tiniest atom; for such an intellect nothing would be uncertain and the future just like the past would be present before its eyes.

But our world today is regrettably not as simple as Laplace had wished for. And it probably was not as predictable even then. Another perhaps more essential form of anticipation is trying to predict what may happen in the future, especially in the environment of the system or company (changing conditions due to competitor initiatives, capricious customer preferences, or social trends). Climate change is only the latest example of how this kind of anticipation has failed, although not for a lack of trying. (The first known scientific presentation of global warming took place in 1895 when the Swedish scientist Svante Arrhenius, gave a talk on "the influence of carbonic acid in the air upon the temperature of the ground" to the Stockholm Physical Society, but the world was about as ready to listen 120 years ago as it is today.) Any such predictions must of course be based on a current understanding of what the situation is, including what information (indicators and trends) is the most important. It also includes trying to guess what other actors or players possibly might do, which leads into the fascinating world of game theory (either classical two-person game theory (von Neuman & Morgenstern, 1944) or the more advanced n-person game theory Rapaport (1970)) in attempts to anticipate what other players, for instance competitors (with an active strategy) in a market are likely to do, how a cancer or a pandemic may develop, how a bushfire or a financial disruption may spread, etc. In each case, anticipation is about potential scenarios and developments, hence about unexampled events (Westrum, 2006) – something that could happen but has never happened before. This makes anticipation both the most important of the four potentials and the most difficult to nurture and assess.

Thoughts about the future

Many have been tempted to say something wise about the future. A few examples are provided here:

Charles Dickens (1812–1870) English novelist

Charles Dickens only said something about the future indirectly, for instance when he let Wilkins Micawber give the following advice to the young David Copperfield:

> "My dear friend Copperfield," said Mr. Micawber, "accidents will occur in the best-regulated families; and in families not regulated by that pervading influence which sanctifies while it enhances the – a – I would say, in short, by the influence of Woman, in the lofty character of Wife, they may be expected with confidence, and must be borne with philosophy."

R.S. Petersen (1882–1949)

"It is difficult to make predictions - – especially about the future." This is often attributed to the Danish physicist Niels Bohr but was actually said by Robert Storm Petersen a well-known (in Denmark at least) artist and jokester

Eric Hoffer (1902–1983), American philosopher

"The unpredictability inherent in human affairs is due largely to the fact that the by-products of a human process are more fateful than the product."

James Baldwin (1924–1987), American author

"No one can possibly know what is about to happen: it is happening, each time, for the first time, for the only time."

Tony Blair (1953-), former English PM

Occasionally even politicians may have something to offer. In a Parliamentary debate, March 14, 2007 (on whether the UK should maintain a nuclear deterrent) the then PM Tony Blair provided the following insight:

> I believe that it is important that we recognize that although it is impossible to predict the future, the one thing that is certain is the uncertainty of it.
>
> (Blair, 2007)

(It is unfortunately philosophically suspect to talk about the certainty of uncertainty. Blair's statement is difficult to disprove, and about as useful as Murphy's Law.)

Norbert Wiener (1894–1954) American mathematician and father of cybernetics

"The present is different from the past and the future is different from the present."

Gro Harlem Brundtland (1939 –) former Norwegian PM

"We can never avoid the dilemma that all our knowledge is of the past, while all our decisions are about the future."

Dr Brundtland's wise words were offered almost a century earlier by Søren Kierkegaard, a Danish philosopher (1813–1855).

Søren Kierkegaard Danish philosopher (1813–1855)

He wrote in Danish: *Livet forstas baglans men ma leves forlans.* Which in English becomes – Life is experienced backwards but must be lived forwards

Ben Bernanke (1953-)

In September 2007 the then former chairman of the US Federal reserve offered the following comment on the market volatility.

> "We are again reminded of the need to maintain appropriate humility in forecasting returns and asset prices."

The hedging fallacy

> "It will get worse before it gets better."

This inane prediction is not attributed to anyone in particular. But it is used almost as often as the safety mantra by politicians and in particular financial and business experts whenever there is turbulence in the economy or in the markets. It is sometimes also used by doctors in futile attempts to soothe patients. What it really means is that the person who speaks does not have any idea about what is happening or what is going to happen; it is a warning to people not to be too optimistic and expect things to get better soon. It is hedging because it other-wise would give people false hope that they afterwards could complain about. It is hedging, but it is also playing it safe. The only case where it does not work is global warming, where it only is going to get worse – unless you are a politician who persistently refuses to accept the facts. Unfortunately this "prediction" does not mean that it, whatever it refers to, will get better after it has become worse; there is no necessary causal relation.

Assessing the potentials

Accepting the argument that the four potentials can be used as meaningful (proxy) indicators for how well a system or an organisation is likely to perform; the obvious next question is how they can be assessed or measured. Lord Kelvin, although speaking about electricity, is famous for offering the following opinion:

> I often say that when you can measure what you are speaking about, and express it in numbers, you know something about it; but when you cannot measure it, when you cannot express it in numbers, your knowledge is of a meager and unsatisfactory kind.

Or, even simpler as "If you cannot measure it, you cannot improve it." The logic is straightforward. If you want to change something then you must be able to determine whether and when a change has taken place (the quality of the feedback) and whether the change was in the right direction. If the aim therefore is to manage – which means change – the potentials, then it is necessary somehow to assess or measure them systematically.

A first suggestion might be to assess the level of each potential as such, similar to assessing the level of safety culture, etc. But it is as futile to ask questions about the level of a potential as it is to ask questions about the "level of resilience" or "level of safety (culture)." Questions such as "what is the potential for monitoring?" or "how large is your potential for learning?" may well elicit an answer but it will not be more meaningful than the question – and it is most certainly not an answer that can be easily used to decide on a relevant response. Each potential can, however, be described as comprising a number of more specific facets or functions that are common to many types of activity and domains. So instead of assessing each potential as a whole, the potentials may better be characterised in terms of the several facets that each represents.

Background and foreground questions

Whenever something is done to change a system, such as introducing new automation, knowledge about the reasons for the change – the rationale or design basis – is needed in order to evaluate how well the change works. In cases where this background information is incomplete or lacking, it is necessary to retrieve and/or supplemented. This can be done by probing the details of the system design as they pertain to the four potentials. The answers to such background questions are essential to develop the foreground questions that are the core of the SPM. Since information about the design rationale can be assumed to remain stable during the change – and presumably also for some time after – it can be considered as background information that only has to be assessed once. Answers to the background questions must be obtained before the SPM is taken into regular use.

Background questions for the potential to respond should be directed at facets such as the justification for the list of events that needed a response, ditto for the list of response, their relevance, the threshold for responses, and the verification of responses – for instance, whether the selection of events and responses is based on tradition, regulator requirements, design basis, experience, expertise, risk assessment, industry standard, or something else. Background questions for the potential to monitor should be directed at facets such as how the indicators and measurements have been selected and how their relevance is established. Other important aspects of monitoring would be the validity of indicators, delays in sampling, how measurements are combined and analysed, organisational role of and support of monitoring, etc. Background questions for the potential to learn should look at the role and importance of learning in the organisation, how learning is resource managed, the balance between reporting and learning, how "lessons learned" should be used – and maintained – by the organisation, and so on. Background questions for the potential to anticipate should examine the purpose and potential value of anticipation, how it fits into a long-term strategy or vision, whether it is an internal or outsourced function, and how it aligns with the organisational culture and values.

Other facets that relate to how the system actually performs will not be stable but are likely to change – and are indeed supposed to change – as a change takes place. Many of the changes are the predictable consequences that constitute the purpose or motivation for the project. However, there will always be some unanticipated changes that could be detrimental to the purpose (Merton, 1936). The SPM uses four sets of questions – called foreground questions – to assess the changes in these facets as a means to manage the change. In contrast to the background questions, the foreground questions should be used repeatedly throughout the project. Foreground questions should refer to what a potential means in practice. Examples of foreground questions for the four potentials are provided in Tables IV.3–6. The questions in the four tables are intentionally generic and they do not refer to a specific process or application. In order to assess the potential to respond it is, for instance, clearly useful to know how serious a condition must be before a response is made or in other words what the threshold is. If it is set too low there will be many "false starts"; if it is set too high, a response may come too late. Yet another question could be to determine whether the required resources are permanently available or whether there will be a delay before a response can begin. By reasoning in this way it is possible to develop a set of questions that together provide a basis for assessing the potential to respond. The same type of analysis can be used to develop sets of questions for the two remaining potentials. In the case of the potential to monitor, it is useful to know the following: the relevance of the indicators or "signals" for the purpose, how often measurements or observations are made, whether there is any delay in interpreting or analysing them, how they are used (as leading or lagging indications, for instance), and how meaningful they are for the sharp end and the blunt end respectively. For the potential to learn, important issues are whether learning has the right focus, whether it is reactive (event-driven) or continuous, how the "lessons learned" are shared and used within a system or system or a company, which priority – and resources – learning is given, etc. Finally, for the potential to anticipate, foreground questions can query the relevance of the strategy, whether it is broad or constrained, whether anticipation takes place often enough, and how the results are used – and by whom. The potential to anticipate differs from the three others – and especially from responding and monitoring – by offering fewer opportunities actually to observe how it is done. Anticipation is rarely a routine undertaking (which itself might be something to ask about), and questions about the potential to anticipate must therefore be based more on theory than on practice. Altogether, the four sets of questions provide the core of a method called Systemic Potentials Management (SPM). The systemic potentials can serve as proxy measures to effectively describe or define what a system or a company ought to do – the performance criteria – and answers given to the questions (the potential performance profile) – consequently represent how well these criteria have been achieved.

Scale invariance

A useful feature of the four potentials is that they are scale invariant. This means that they can be used to characterise performance on all levels – from the sharp end to the blunt end and everywhere in between. The potentials to respond, to monitor (to keep an eye on things), to learn, and to anticipate are relevant for operational first-order production processes as well as for second-order strategic processes – for people regardless of where their place of work is, of what type their work is, and which responsibilities they have for the management of the department or unit, but the potentials also apply to law makers and regulators. It is important everywhere to be able to respond, monitor, learn, and anticipate.

The relative importance of the four potentials does, of course, to some extent depend on the nature of the activity or focus. If the main activity is to ensure a smooth movement or flow of something – vehicles on land, on sea, or in the air; raw materials; or finished products, then the potentials to monitor and to respond may be more important than the potentials to learn and anticipate. Conversely, for long-term planning and management of traffic or production, learning and anticipation may be more important than responding and monitoring. Nevertheless, the potentials can be used to keep track of how well a system is able to perform at all levels, regardless of the scope and the temporal assessing how well a system is able to perform at all levels, regardless of the scope and the temporal dynamics. – whether it is something that happens now and in the foreseeable future or something that evolves over an extended period of time.

It is not reasonable to prescribe an ideal balance or ratio among the four potentials that is independent of the domain. For a fire brigade, for instance, it is more important to be able to respond than to anticipate, whereas for a gadget manufacturer the potential to anticipate (for instance, customer preferences or social trends) may be just as important as the potential to respond. Resilience engineering nevertheless makes it clear (Hollnagel, 2009) that it is necessary for a system or a company to have all four potentials to some extent, in order to be able to perform as required in expected and unexpected situations alike (as the definition of resilient performance goes). All systems usually put some effort into the potential to respond, because it is so obviously important. Most also put some effort into the potential to learn, although it often is in a perfunctory manner. Fewer systems or companies make a sustained effort to monitor, particularly if there has been a long period of stability when "nothing" has happened. Finally, very few systems or companies put any serious effort into the potential to anticipate.

The four potentials as weak signals

As explained earlier, the weak signals include the spatial and temporal patterns that people recognise and use to manage their work. In the case of intangible systems – as in change management in organisations – temporal patterns are difficult to define and recognise because they cover time spans too long for humans to comprehend. A possible solution is to apply something similar to a time-lapse approach as it is used in photography and film, except that in this case the interval between measurements may have to be in the order of months or years rather than minutes, hours, or days. If meaningful signals could be defined, this would make it possible (and sensible) to look for temporal patterns in the ways a system or a company perform, hence to manage that performance safely in real time.

What constitutes a meaningful signal will of course depend on the nature and characteristics of the process being managed. In line with the previous arguments for a unified approach, the signals should be relevant for multiple foci or priorities, rather than particular to just one of them. Given that it can be quite difficult to find specific signals in a system's performance (just think of all the efforts that have been spent to find good indicators of safety, quality, etc.), it may be better to look at what lies behind, at what determines or shapes performance and to use this as proxy measures. The patterns in performance are first of all determined by the regularity (patterns) of what happens in the surroundings (the context in which the system or company operates), hence by the nature of the process(es) being controlled. But performance patterns are also determined by general characteristics of how a system or a company responds, monitors, learns, and anticipates, similar to the way observable organisational performance (artefacts and behaviours) depends on unobservable espoused values and basic underlying assumptions (as in the organisational culture described by Schein, 1992) in the case of slowly developing processes, such as safety, quality, etc. The potentials – or rather the assessed status of the potentials – may therefore serve as a source of information (or weal proxy signals) that can be used in the management of changes, as long as it is possible to assess them systematically and reliably.

Developing detailed assessment questions

The SPM is intended as a tool for managing how well a system or an organisation performs and how changes are implemented. Since the generic questions (provided in Table II.5 do not refer to any particular domain or process, the first step is to develop questions that are diagnostic for the organisation or change being considered. The questions should be about operational aspects, about how something is being done or how something happens rather than about how someone thinks about it or likes/dislikes it. The generic foreground questions (Tables IV.5) can serve as a starting point, but only questions that are acknowledged as relevant for the actual purpose should be used. The generic questions can and must obviously be supplemented by additional questions based on knowledge about the organisation and the nature of the intended change. The questions should be *diagnostic*, for example directed at problems or issues that are known to be relevant and therefore meaningful to assess, even if they have been "missed" by the generic questions.

When a question is asked it is important that it is meaningful but also that the answers are concrete and practical. In the context of the four potentials the questions should be *formative*, so that the answers easily can serve as a starting point for proposing specific actions or changes to improve – or maintain – the conditions, keeping in mind that the potentials are not independent of each other.

To illustrate a question, consider for instance the following statement used as a question:

"The organisation learns from reported events."

The answers to a statement would typically be either assenting or dissenting. Such answers would indicate the general attitude towards organisational learning among the responders, but the answers would not be directly applicable to develop an effective intervention or improvement. If the answer is dissenting, it would clearly be necessary to do something. But the statement is not specific enough to help with that. The statement could, however, also be expressed as an assertion as follows:

"There are sufficient resources to write reports."

Or

"Submitted reports are being investigated sufficiently."

In both cases it is far easier to think of what to do, in the case of an assenting answer. Responses should serve to maintain conditions and in the case of a dissenting answer they should serve to do something that effectively will improve them.

Yet another example is the second question in Table IV.4, which is

"Are the set of indicators regularly revised?"

If the answers show this is to be the case, meaning that the respondents agree that indicators are revised regularly and properly, then that is a good basis for deciding what to do, specifically how to sustain the condition, even though the choice of means will require detailed knowledge about how the organisation in question works rather than a generic, standardised solution.

How to formulate the questions

The foreground questions shown in Tables IV.3–6 have all been formulated in the same manner, but it is obviously possible to formulate questions in many different ways, for instance as statements that respondents can agree or disagree with.

In either case the answers can be binary ("Yes" or "No") or graduated by using a Likert scale. Questions can also be open-ended or closed-ended, negative inviting dissent ("Indicators are not revised regularly and properly") or positive inviting consent ("Submitted reports are being investigated sufficiently"). There is a rich literature in social psychology about how to formulate questions and conduct surveys and interviews. Questions should generally be descriptive (referring to what goes on or happens) rather than relational (speculating about how something relates or compares to something else) or causal (speculating about what-if relations). Questions should be formulated so that they are meaningful to respondents and easy to answer, because they refer either to something that is part of the respondents' competence or experience or something that the respondent is knowledgeable about in general. *But most importantly, a question should never be asked unless you are sure you know why you need the answer and also know how to make use of it!*

The purpose of the systemic potentials is to support the management of organisational change rather than to establish a temporary benchmark. It is therefore essential that the questions are administered not just once but at regular intervals. The time between surveys depends on the nature and pace of the process or change being managed and requires a good understanding of how rapidly (or how slowly) changes take place, just as the frame rate in time-lapse photography is determined by the natural speed of what is being recorded (birds building a nest or flowers growing). The answers to the questions provide information about an organisation's "position," and that information must be updated as the situation changes. (Consider, for instance, how many years it takes to change a safety culture or the espoused values of an organisation.) How frequently the questions should be used depends on the nature of the change, but once every 3–6 months might be a reasonable interval for many organisational changes (yet again do not assume that this is a norm). The frequency should match the (assumed) rate of change being considered – either a change being deliberately made or the changes/fluctuations in external conditions that require management interventions. The number of times the SPM should be used or the overall duration also depends on the nature of the change, but it may easily be in the order of years rather than months, although this should not be taken as a norm. Trying to determine the answer will itself help to build a better understanding of what goes on and what plan, do, and study therefore need to address.

Since the assessments are intended to support the management of change, it is important that they are reliable. This can be ensured by having, as far as possible, a stable group of respondents. They must clearly be people who are directly involved with the functions that are being managed, but it can also be useful to have people in different roles and organisational positions – or even outside experts as a way of calibrating or checking for potential biases in the answers. The number of respondents should not be too large and it should also be easy, quick, and convenient to answer the questions. With present-day technology a web-based survey might be an attractive option.

Interpreting the answers

Because the SPM is a tool for change management, the answers cannot simply be compared to a standard or external reference but only to other earlier answers from the same organisation or from departments of the same organisation that can be assumed to be very similar (e.g., between two drilling platforms or two service units). (And also from the same respondents if possible.) By comparing answers from repeated applications of the question sets, the answer profiles will show how an organisation changes over time. Quite generally any socio-technical system should only be compared to itself, since any two socio-technical systems will be different, comparing them – or even worse, comparing them to a standard – will therefore be folly, which perhaps is why it is normally done. The only relevant comparison will therefore be a historical one to the organisation itself – over time!

A convenient way to construct a profile is to use the so-called radar charts or nets that are part of most analytics packages (and spreadsheets). If the answers are given on a Likert scale it is simple to produce the corresponding graphics. Otherwise, some kind of transformation is needed. An example of what the answers may look like is shown in Figure IV.15, in this case for facets of the potential to respond. The values are for the purpose of illustration only and do not represent an actual sample. By comparing the profiles from months 4 and 8, it is easy to see for which facets the ratings have become more acceptable, which have stayed the same, and which have become less acceptable. In this way, the two profiles show the change in "position" of the organisation with regard to the potential

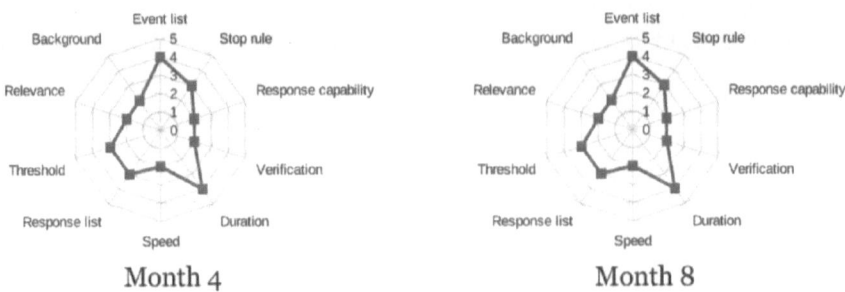

Figure IV.15 Systemic profiles for the potential to respond

to respond. Similar profiles should be made for the other potentials. Although it may be tempting or perhaps even irresistible, the ratings of the individual facets should not be combined to produce a single score for any of the potentials.

Visualising temporal patterns

The radar chart is a convenient way to visualise a potentially complicated relation but not actually the weak signal we need (Figure IV.15). The radar chart is analogous to a frame in time-lapse photography and the weak signal is the temporal pattern – i.e., the changes in the shapes of the polygons – that appears when multiple profiles or frames are seen together, as a kind of movie. It is of course also useful to look more closely at each profile by itself, to determine whether the ratings of the facets are as expected and as intended or required. If they are not, it may be because the frequency was too high (effects from earlier interventions may not have manifested themselves yet) or because the chosen interventions did not work as intended.

The profiles of the systemic potentials not only provide a practical way to determine what the situation or position is, but they can also be used to represent what the goal or target should be. The goal or purpose of an organisational change is usually given verbally or at best as a specific numerical value, such as zero accidents, an X percent reduction in reported events, or a Y percent increase in quality at a certain point in time. But if a goal is thought of in terms of specific targets for each facet, it may easily be transformed into a profile for any or all of the potentials. One way of showing that is illustrated by Figure IV.16. Here the outer profile (dashed line) represents the goal, while the inner profile (full line) shows the situation at a specific point in time.

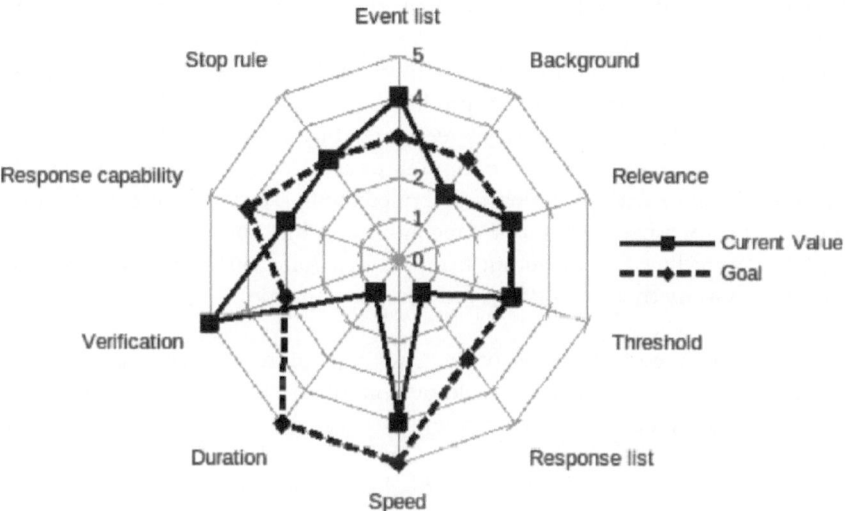

Figure IV.16 A goal shown as a profile

The goal could, of course, also have been shown simply as a regular polygon with the maximum scoring of answers, in this case 5. That would correspond to a statement such as "the potential to respond should be as high as possible." But by differentiating among the various facets of the potential to respond, it becomes possible to express the goal in a more nuanced manner to signify that some facets, in this case speed of response, are considered more important than others.

Dependencies or couplings

A radar chart may reveal changes to facets due to underlying dependencies that have not been correctly understood – meaning that the underlying model is inadequate in the cybernetic sense discussed previously. When it comes to specifying the means, i.e., the detailed planning and implementation of interventions to bring about specific changes, it is essential to keep in mind both that the four potentials are mutually coupled but also that the facets – the more detailed functions – will depend on each other. While it may be tempting to address the potentials or their facets one by one, it is clearly not advisable. Even the brief description of the potentials given here makes it clear that they depend on each other – for instance in the way that responding depends on monitoring. Recognising and understanding these dependencies is not only helpful in formulating the specific questions but will also be indispensable in choosing what interventions or modifications to make. The same, of course, goes for the facets of the four potentials.

Understanding the dependencies among the four potentials – or even better, the dependencies among the facets of the potentials – requires a detailed model of how an organisation functions. While it is not possible here to explain how such a detailed model can be developed, one approach could be a functional resonance analysis using the functional resonance analysis method (FRAM; Hollnagel, 2012). Figure IV.17 illustrates how this might look for a generic rather than a specific organisation. Suffice it to say, it is important to resist using a "one problem – one solution" approach in any kind of systemic management, whether the focus is safety, quality, productivity, resilience, or something else.

In Figure IV.17, each potential is represented by a function. The potential to monitor corresponds to the function monitor and so on. In accordance with the FRAM formalism, each function is described by six aspects, represented by their initials. The six aspects are input, precondition, time, control, and resource; all aspects are produced as an output from another function. The functional resonance analysis method is based on four simple principles:

- The principle of *equivalence* of successes and failures (which basically means that acceptable and unacceptable outcomes are the result of the same kind of performance (Mach's principle)).
- The principle of *approximate adjustments* (which basically says that people always adjust how they work to fit the conditions as they see them and understand them).
- The principle of *emergence*, which replaces linear causality.

- The principle of *functional resonance*, which explains how non–linear effects (meaning effects of disproportionate magnitude) may happen, using the analogy with the resonance as it is known from classical physics.

In the generic model seen in Figure IV.17, the input to anticipate is the lessons learned, which is one of the outputs from learning. The lessons learned are also used to control the monitoring function. Another output from learning is the sampling frequency, which is coupled to the time aspect of monitoring. The model shown in Figure IV.17 is generic, but a specific model should be developed before the SPM is being used for a system or a company. The model will be of help in deciding on interventions but may also be useful in formulating the detailed questions. Developing a model is a skill that is quickly learned and the work is made easier by the open-source software tool known as the FRAM Model Visualiser (FMV), which was also used to develop the model seen in Figure IV.17.

The SPM model shown in Figure IV.17 is a proper systemic model for the following reasons:

It does not assume that any of the four main functions can operate independently of the others.

Neither does it assume that a function works in the same way when it is examined singly and when it is part of the whole.

And it is fully acknowledged that the overall performance of the system/model is subject to feedback loops and non–linear interactions. (That is indeed the essence of resonance and emergence.)

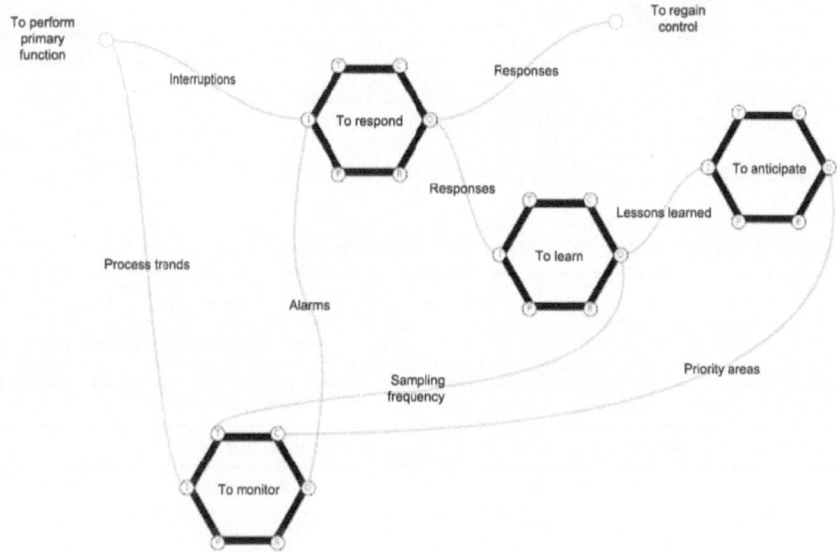

Figure IV.17 Couplings among the four potentials, for a generic system

Table IV.3 Possible foreground facets for the potential to respond

The set of events for which responses exist is adequate.
The set of events is regularly reviewed/revised.
The frequency of such revisions is adequate.
It is clear when a response should be given.
Responses are simple to make and do not require ad hoc choices.
The prescribed responses are appropriate for the situations where they are needed.
Responses can be started/initiated fast enough.
Responses can be fully implemented fast enough.
Effective responses can be sustained for long enough.
There are sufficient resources (people, equipment, materials) to ensure response readiness.
The resources are maintained at an adequate level.
The readiness to respond is regularly assessed and maintained.

Table IV.4 Possible foreground facets of the potential to monitor

The set of indicators is adequate for the purpose.
Any revision makes use of practical experience.
Monitoring is recognised as an essential part of the organisation's performance.
The set of indicators is regularly revised.
The ratio of "leading" to "lagging" indicators is appropriate.
The basis for the indicators is well articulated and easy to understand.
Indicators are checked with sufficient frequency (continuously, regularly, every now and then).
The indicators are directly meaningful and do not require additional interpretation.
Monitoring is recognised as an important part of the organisation's performance.

Table IV.5 Possible foreground facets of the potential to learn

The organisation tries to learn both from successes (work that goes well) and from failures (work that did not go well).
Event reports are meaningful and easy to understand.
Event reports contain sufficient details and data.
There are well-defined procedures for reporting, analysis, and learning.
There is adequate formal training and organisational support for reporting, analysis, and learning.
Learning is continuous rather than event-driven.

The organisation tries to learn both from successes (work that goes well) and from failures (work that did not go well).
Adequate resources are allocated to reporting, analysis and learning.
The delay between reporting and analysis results (lessons learned) is acceptable.
The outcomes from learning are communicated effectively throughout the organisation.
The lessons learned are directed at the right level (for instance, individuals, groups, departments).
The "lessons learned" are properly implemented (as regulations, procedures, norms, training, instructions, redesign, reorganisation, etc.).

Table IV.6 Possible foreground facets of the potential to anticipate

Future threats and opportunities are assessed with sufficient frequency.
Information about future events is communicated or shared within the organisation.
The organisation has a clearly formulated vision for the future.
The organisation is clearly concerned about what could happen far ahead.
Humility is a shared organisational virtue.

- The questions must be *specific* and refer to the system or company in question. The questions should be about operational aspects, about how something is being done rather than about how someone thinks about it or likes/dislikes it. The generic foreground questions can serve as a starting point for developing application-specific questions but must never be used as they are. Only questions that are considered relevant for the actual purpose should be used. It is advisable that a team of experienced people representing both the sharp end and the blunt end develop four sets of specific questions, based on the generic questions. This discussion alone can provide valuable insights on how the organisation functions.
- The questions must also be *diagnostic* and concerned with problems or issues that are known to be relevant (perhaps representing past problems) and therefore are meaningful to assess. The questions should preferably refer to recognisable activities or functions, which can be assessed in terms of whether they are adequately performed – as operations rather than as subjective impressions. If the generic questions seem to miss known important issues, additional diagnostic questions should be added as needed.
- The questions must finally be *formative* so that the answers either can be used directly to choose an appropriate intervention or serve as a starting point for proposing specific actions or changes to improve – or maintain – the conditions revealed by the surveys. It is essential to keep in mind that the potentials are not independent of each other in order to keep unanticipated consequences

to a minimum. The understanding of the ways in which the potentials depend on each other is equivalent to a functional model of the system or company works, and this understanding or model can be useful both in deciding what to do and in predicting the effects of the chosen action(s). Questions should be as concrete and operational as possible.

How to formulate the questions

- To illustrate the importance of how the questions are formulated, consider the following question addressing the potential to learn:
- "The organisation learns from reported events."
- This way of formulating the question invites a binary YES or NO answer, which is actually not very useful for deciding what to do. Answers would typically be either assenting or dissenting. This would indicate the general attitude towards organisational learning among the responders; the question may well be diagnostic but it is not formative and therefore not very useful to develop an effective intervention. The question simply invites people to answer either "Yes" or "No." (Try to think of what you would do, in either case.) The potential to learn could, however, also be probed in the following more specific way:
- "There are sufficient resources to write reports."
- In this case even a Yes/No answer could suggest some kind of response, because the question is both diagnostic and formative.
- Or, even better:
- "Submitted reports are being investigated sufficiently."

In this case it is far easier to think of what to do, even if the answer again is a simple "Yes" or "No."

Yet another example is the second question in Table IV.4, which is

"Any revision makes use of practical experience."

If the answers show this not to be the case, for instance if the respondents disagree that indicators are revised based on practical experience, then that is a starting point for deciding what to do, even though the choice of means requires detailed knowledge about how the organisation in question actually works rather than off-the-shelf solutions. The foreground questions shown in Tables 4.3–6 have all been formulated in the same manner, but it is obviously possible to formulate questions in many different ways, for instance as statements that respondents can agree or disagree with:

"We revisit and revise our list of events and action plans on a systematic basis."

or

"The period covered by the lagging indicators is appropriate."

or

"The employees are being motivated to write reports."

Once more the golden rule, in preparing the sets of questions and in formulating the detailed questions, is that a question should never be asked unless there is a clear operational reason for it.

Administering the questions

The purpose of the systemic potentials is to support the management of organisational change rather than to establish a temporary benchmark or a basis for comparing to some standard. The surveys are a tool to get information about the current position, since all organisations as socio-technical systems are unique, even when it is attempted to make them identical copies of something (think fast food chains, but even if even if two fast food restaurants are identical, the people who work there are not). An organisation can anyway only ever be compared to itself over time. But that is not necessarily a limitation; it is precisely what is needed for change management. It is therefore essential that the questions are administered not just once but at regular intervals. The answers to the questions provide a snapshot, so to speak, of an organisation's state or "position" at that time, and that information must be updated as the situation changes. (Consider, for instance, how long it will take to change the safety culture of an organisation. The problem is that few people have a clear idea about that.) How frequently the questions should be used depends on the nature of the change, but once every 3–6 months could be a reasonable interval for many organisational changes. If the interval is too short the results of interventions may not have had time to become manifest, and if the interval is too long, other unknown and uncontrolled factors may have played a role. The frequency should match the (estimated) rate of change being considered – either a change being deliberately made or the changes/fluctuations in external conditions that require management. Whatever the estimate is it will have to be calibrated as part of the process, and this must be realised from the start. The number of times the SPM should be used or the overall duration also depends on the nature of the change, but for an organisational change it may easily be in the order of years rather than months. Sustaining motivation and financial support may therefore be other issues to consider.

Since the assessments are intended to support the management of change, it is important that they are reliable. This can be ensured by having, as far as possible, a stable group of respondents. They must clearly be people who are directly involved with the functions that are affected by the intended changes, but it can also be useful to include people in different roles and organisational positions in the surveys – or even outside experts as a way of checking for potential in-house biases in the answers. The number of respondents should not be too large and it should also be easy, quick, and convenient to answer the questions. With the current technology it is clearly something that could be web-based.

Management epilogue

After having established and characterised the current state, the safety legacy (Part I), and complexity (Part II) and recognised the futility of accident investigation (Part III), Part IV has considered the necessary consequences, such as that management must be concerned less with *what* happens – where management has been based on counting certain categories of outcomes – and more with *how* it happens – where management must be based on assessing whether the qualities or potentials for the desired performance have the presence that is required, hence with a focus more on what does not happen, on the second-order (cultural) activities that provide the foundation and support for the first-order (production) activities dynamic non-events can be very slow. And unlike physical processes such as a production line or a building site, there are no well-defined critical functions for organisations, hence no direct measurements available. In their place management must rely on proxy measurements, but it is essential that there is an articulated relation between the proxy measures and the performance that is sought. Managing organisational changes addresses second-order activities that are at least an order of magnitude slower than the first-order activities on which most management thinking and techniques are based, and managing cultural changes is a third-order activity that is even slower.

Two problems: the first problem is that there are few, if actually any, well-defined measurements. The second problem is that organisational change is slow though no one knows precisely how much time it takes, and few seem to have cared. Anecdotal evidence suggests something on the order of 10–15 years and changes that take so long can be difficult to detect and/or manage. The real time of our mental imagery cannot easily be reconciled with the real time of an organisational change

Since there are no direct measurements to be made, there are no strong signals in the traditional sense, but in their place there are a number of weak signals. These are typically temporal patterns that may be difficult to recognise precisely because they must be seen over longer periods of time. The Systemic Performance Monitoring based on the principles of resilience engineering offers a practical solution to all these problems by using systematic assessments of the four potentials as proxy measures. When made regularly the results can be used as the basis for a simple graphical representation where patterns (or weak signals) are easy to detect.

Managing safely thus relies on indirect performance measures that represent the **presence** of safety, in contrast to safety management which relies on measures of what happens in terms of acceptable and especially non-acceptable outcomes that mainly represent the **absence** of safety.

V Coda: connecting the dots

Coda prologue

Accepting the problems of the safety legacy and vision zero, the obvious question is, what should be put in their place? Managing safely must, of course, have an articulated conceptual and methodological basis. It is necessary to consider the practical and theoretical consequences of the differences between safety management and managing safely, as well as several other issues mentioned in the previous pages.

How about Safety-III?

Some readers may at this point wonder why this book has not been about Safety-III. The reason is simply that Safety-III is a meaningless concept.

The naming of Safety-I and Safety-II was inspired by Ed Dougherty's (1990) provocative but long forgotten plea for a second generation in Human Reliability Analysis (HRA), The problem in putting forward the terms "Safety-I" and "Safety-II" in Hollnagel (2014a) was that they, despite the deliberate use of Roman rather than Arabic numerals, might be seen as representing an arithmetic progression. If so, it would be deceptively easy to "conclude" that the sequence could be continued to a Safety-III, Safety-IV, and ultimately some higher form of Safety-N. But such "inference" is based on a fundamental misunderstanding of the terms and of the intention behind the proposal of Safety-II, as Safety-I and Safety-II represent a rhetorical rather than a numerical relationship. (It also conveniently overlooks the deliberate use of Roman rather than Arabic numerals, but the temptation was apparently irresistible for several, e.g., Aven (2022), Leveson (2020), and Cooper (2022), the last even failing to distinguish between Safety-II and "safety differently" (Dekker, 2015).) The terms Safety-I and Safety-II were chosen to juxtapose alternative views of safety or rather different ways to understand and manage how socio-technical systems function – in particular those we call a Complex Adaptive System (CAS; cf., Part II)

DOI: 10.4324/9781032664729-5

The possibility of this misunderstanding was anticipated even when the terms were first proposed. The main arguments from Hollnagel (2014a) are worth repeating here:

> Since Safety-II represents a logical extension of Safety-I, it may well be asked whether there will not someday be a Safety-III. In order to answer that, it is necessary to keep in mind that Safety-I and Safety-II differ in their focus and therefore ultimately in their ontology. The focus of Safety-I is on things that go wrong, and the corresponding efforts [naturally] aim to reduce the number of things that go wrong. The focus of Safety-II is on things that go well, and the corresponding efforts [naturally] aim to increase the number of things that go well.

Safety-II thus represents both a different focus and a different way of looking at what happens and how it happens. Doing this will, of course, require practices that are different from those that are commonly used today. But a number of these practices already existed, either in principle or in practice, as described in Hollnagel (2014b) and could easily be taken into use.

> It would, of course, also be necessary to develop new methods and techniques to make it possible to deal more effectively with what goes well and which are able in particular to describe, analyse, and represent the ubiquitous performance adjustments.
>
> (Hollnagel, 2014b, p. 177)

Among the practices mentioned in Hollnagel (2014b) Chapter 8 were "looking at work that goes well," "breadth before depth," and "appreciate and manage performance variability."

If the way ahead is a combination of current practices of Safety-I with the complementary – and to some extent novel – practices of Safety-II, then where does that leave Safety-III? It has been suggested that Safety-III "simply" stands for the combination of existing and novel practices. But the combination of practices is not against the idea of Safety-II, which is intended as a complement to Safety-I rather than as a replacement of it. Neither does the suggestion of a possible "Safety-III" offer a new understanding of safety, a new ontology, in the way that Safety-II does, and it is for that reason alone not necessary. It can, of course, not be ruled out that in some years' time there may come a proposal for understanding safety with a definition and ontology of its own that is different from both Safety-I and Safety-II. It may also happen that the very concept of safety is gradually dissolved, at least in the way it is used currently; others will hopefully take up the mantle, as something distinctively different from it, for example quality, productivity, efficiency, etc. (and Hollnagel, (2020) was a feeble attempt to do that, although not clearly anticipated in 2014). If and when that happens the result will not be a Safety-III but rather a whole new concept such as *synesis* (Hollnagel, 2020). So, while Safety-II by no means should be seen as the end of the road in the efforts to ensure that socio-technical habitats function as well as we need them to, it may well be the end of the road of safety as a concept in its own right. (Hollnagel, 2014, p. 178).

The argument can be summarised like this: Safety-I represents the concern for managing events with unacceptable outcomes. This follows the Heinrich dogma by trying to explain how things go wrong in order to prevent any recurrences. The focus on work that goes wrong in practice excludes everything else. Safety-II represents a concern for managing how things happen regardless of whether the outcomes are acceptable or unacceptable but especially looks at work that goes well. This is done by trying to understand how work goes well in order to facilitate acceptable outcomes but also to dampen or prevent unacceptable outcomes.

The arguments can also be shown graphically (Figure V.1). If we assume that outcomes follow a normal distribution, which they don't necessarily, the focus of Safety-I is on rare events with unacceptable consequences as shown in Figure V.1.

Safety-I looks at outcomes that only happen infrequently and that are unwanted (the events with unacceptable outcomes or work that goes wrong, hence the low probability outcomes at the left tail of the distribution). Safety-II looks at all events regardless of their outcomes but in particular at the events that occur frequently and lead to the expected outcomes and which therefore are seen as "normal," occupying the middle of the distribution. These are the dynamic non-events that Weick (1987) wrote about. Since Safety-II is concerned with everything that happens (and not just with work that goes well or the positive surprises, corresponding to the infrequent unexpected successes at the right tail of the distribution), there is nothing else to look at. And since Safety-II looks at all outcomes regardless of whether they are acceptable or unacceptable, there is no other way of looking at them. Safety-II, of course, has a bias towards frequent events with acceptable outcomes but only because these traditionally have been neglected or excluded as having little or no interest.

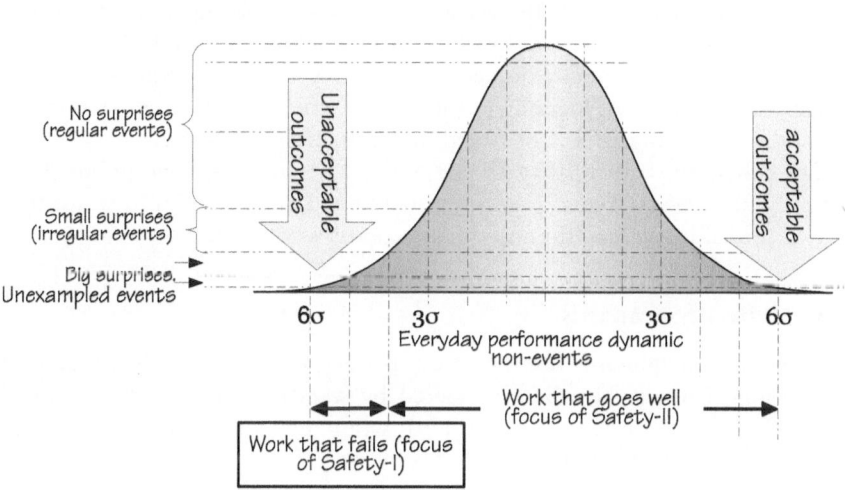

Figure V.1 Different foci of Safety-I and Safety-II

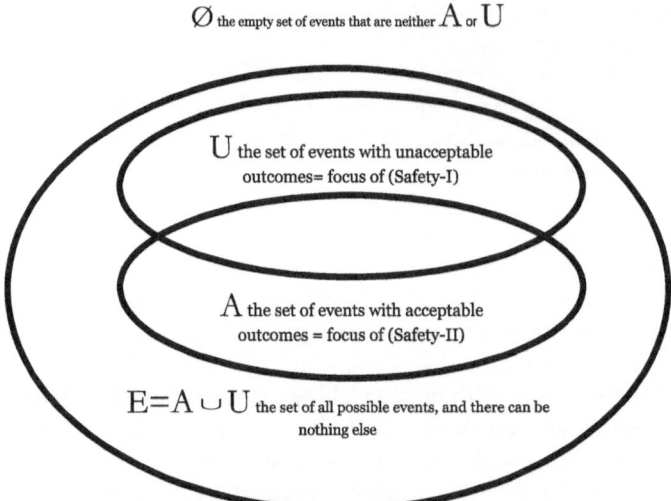

Figure V.2 Venn diagram showing why Safety-III is logically impossible

The argument can also be made more formally, as illustrated by Figure V.2

Consider the set of all events, U, where the outcome is seen as unacceptable. Then consider the set of all events, A, where the outcome is seen as acceptable. Anything that can happen must be a member of the union of the two sets, $E = A \cup B$. Beyond that is only the empty set $\varnothing = A \cup B$.

For an engineering view of the safety of socio-technical systems, the focus is limited to U and the approach is that of Safety-I. For a systemic view the focus is on E and the approach is that of Safety-II.

There could, theoretically, be a study of only A, but it would not make much sense since it would exclude the events with unacceptable outcomes, for example U. Since Safety-II is the study of E, it does by definition include the concerns of Safety-I. There is therefore no need for a "Safety-III" or of a Safety-IV or any higher order, and it not logically possible since there cannot be any events that are not members of either A or U and therefore of their union E.

The conclusion is that Safety-III not only is a meaningless concept but also is logically unnecessary. Yet, as previously mentioned, such niceties have not prevented several individuals from speculating about a Safety-III.

The allure of standards

The ten commandments illustrate how people and societies always, at least since biblical times, have been attracted by standards but also how limited their effects are. Efforts to improve both safety and quality in practice often rely on standards, and it is generally taken for granted as it must be that the standards themselves are complete and correct. Since following the guidance provided by the standards to the letter is assumed to compensate for human shortcomings and performance variability and result in work that is correct and flawless as long as work and workers comply with the standards, there is usually an insistence on compliance to standards, where non-compliance is a frequently used explanation when something has gone wrong even to the extent that there is a special taxonomy for various forms of non-compliance (formerly called violations; Table V.1). The attractiveness of standards for safety management is that they are assumed to be ready-made solutions that provide the information that is necessary and sufficient for work to go well as long as the situations that were assumed or imagined by the people who developed the standard match the actual situations where the standards are applied. According to this logic the solution is compliance; an added benefit is that this reduces the efforts required from people in managerial position. Instead of having to understand what goes on they can just put a checkmark in a box on a list. The bureaucratic insistence on strict compliance is therefore a myopic and ineffective solution in the long run.

Comment: The interesting question is, of course, how often do these trade-offs play a role in work that goes well? And why are they not noticed?

Table V.1 Categories of non-compliance (Hudson et al., 2008)

Type of non-compliance	*Description/example*
Unintended forms of non-compliance	
Unintended understanding failure	when people have a different understanding of what the procedure is and what they have to do
Unintended awareness failure	when people are unaware of the existence of a rule or procedure and therefore work without any reference to it
Intended forms of non-compliance	
Situational non-compliance	when the situation makes it impossible to do the job and be compliant, e.g., because of insufficient time or resources
Optimising non-compliance for company benefit	individuals take short-cuts believing that this will achieve what they believe the company – and their managers – really want
Optimising non-compliance for personal benefit	short-cuts taken to achieve purely personal goals
Exceptional non-compliance	short-cuts if the formal procedures is too difficult to follow under specific, usually novel, circumstances

Scientific management

Standards are attractive because if something follows a standard in the way it is made and in how it works, uncertainty is minimised. The impetus came from the use of standardisation in manufacturing and assembly of products (Womack et al., 1990). A standard is something like a weight, a measure, or an instrument by which the accuracy of others is determined, hence a reference of some kind so you can rely on parts being what they should be and to function exactly as expected. (Norbert Wiener, however, once commented that the problem with computers is that they do exactly what we tell them to do and not what we thought we told them to do.) A standard supports the illusion that you know exactly what happens and what will happen. A standard ensures that the component you use will fit perfectly with other components, like nuts and bolts. If it worked for physical objects, then why not try it also for humans, who for centuries have been likened by philosophers to sophisticated machines (de La Mettrie, 1996, org. 1747) Standards are attractive because they increase efficiency and promise to overcome the shortcomings and imperfections of humans – exactly what human factors engineering had set out to do (Fitts et al., 1951). A standard answers the question "what is the best way of doing something?" that was posed by Frederick Winslow Taylor (1856–1915), the father of scientific management, a system that today is better known as Taylorism. Taylor was also the first efficiency expert and, together with Frank Gilbreth, a pioneer of time and motion study. "In the past the man has been first. In the future the System will be first," he predicted boldly and accurately. His ambition was a clockwork world of tasks timed to seconds (Kanigel, 2005). And this attitude still for many people determines how they think about work management Although standards are meaningful in the physical world and in relation to physical objects, it is more dodgy when it comes to individuals and organisations and their functioning rather than their physical characteristics. The military may well train or drill soldiers until they all act and react in the same way, and this is useful for marching in formation, but this approach is not suitable for fighting a modern war or for work in general. scientific management unsuccessfully tried to do the same for workers. as is clear from the descriptions of the four principles that provide the basis for scientific management (cf., Taylor, 1911, p. 14).

- **First principle**: Develop a science of work with rigid rules for each motion of every person and the perfection and standardisation of all implements and working conditions.
- **Second principle**: Carefully select and train workers into first-class people and eliminate all people who refuse to or are unable to adopt the best methods.
- **Third principle**: Bring first-class workers and the science of working together, through the constant help and watchfulness of the management and through paying each person a large daily bonus for working fast and doing what he is told to do.

- **Fourth principle**: Reward/respond to work quality directly. High pay for success – each worker should be rewarded when he accomplishes his task. Loss in case of failure – when a worker fails, he should know that he would share the loss. This last part, for some reason, never became popular.

Standardisation is really an efficiency-thoroughness trade-off, where the thoroughness has been delegated to other people in a different location and at a different time and this trade-off is used as justification for being efficient rather than thorough in the situation where the standard is applied. Without standardisation of components something like an assembly line would not be possible. The downside of standardisation is that people may assume that it can simply be used without having to think about the situation, as long as the instructions and guidance are followed to the letter. Such blind following requires that the conditions that the developers of the standard imagined or had in mind when the standard was produced perfectly match the conditions that exist when it is to be used. But we know that is rarely if ever the case. Standard writers are also subject to the limitations of requisite imagination.

The allure of standards can be seen in the way that people look for and use, for example, ISO safety standards, but it is a grave mistake to assume that following a standard to the letter guarantees that all will be well; ironically the opposite is more likely to be the case because the standard encourages a lack of thoroughness. There can be standards for physical components but not for a Safety Management System. Standardisation creates the false illusion that events and developments are perfectly predictable, hence that there will be no surprises. But for this illusion of standardisation to work it is necessary that the system works with perfect precision and that there is no variability in the way that system components, including people at either the sharp or the blunt end, do their work, but we know that this is not the case and that it never can be so, especially if the system is a CAS (where the A in CAS means adaptive, which is incompatible with standardisation). Assuming that human performance can be standardised, individually or collectively, is an unsound illusion. There are, so far, no standardised ways of applying a standard. If there were, we would not need people to do it but could delegate all work to machines, which would not even have to be very sophisticated. And since this crucial precondition for standardisation cannot be upheld, the overall hope that standardisation will solve the problems of safety management must be abandoned. A standard is not a carte blanche to stop thinking and being thorough, even though it appeals to the many who seem to think so.

Why safely rather than safety?

There are several reasons why it is preferable to manage safely rather than employ safety management, first because it is far more interesting from both a theoretical and a practical perspective to understand how work goes well, which it does nearly all the time, than to understand why it occasionally – but rarely – fails. It is interesting from a practical perspective because such understanding contributes directly to increased productivity. And it is interesting from a theoretical perspective because it exposes the limitations of the safety legacy and of thinking in terms of simple linear causality. We have since Heinrich (1931) been conditioned to believe that each failure has a single and unique cause, the extreme case being the assumption of the root cause. But no one with even a little experience from practical work would even for a moment propose the corresponding hypothesis that successful outcomes have just a single cause (The US Institute for Healthcare Improvement (IHI) has nonetheless proposed a simple linear Success Cause Analysis (SCA), which uses root cause analysis methodology to understand the factors that contribute to favorable outcomes instead of adverse events (sic!)). Most sensible industries understand that successful (acceptable) outcomes emerge from an intricate and fortuitous combination of factors, conditions, and circumstances that we are unable to predict and it is only reasonable to assume that the same also is the case when something fails, except the combination then is unfortunate rather than fortuitous. We must therefore willy-nilly think about what happens in a different way, and this may be useful also when we try to understand why something has failed or gone wrong.(cf. Russell, 1913).

Managing safely also increases the synergy with quality and productivity as envisaged by the concept of synesis (Hollnagel, 2020).

Safety can anyway not be managed as such because it is defined by its absence rather than its presence (Reason, 2000, p. 3). It is in practice only possible to manage something that is present and which therefore can be influenced in the way it happens. (Until a working time machine becomes available, we cannot change what happened in the past, and even then we must be extremely careful, as many science fiction writers have warned us.) In the change from safety to safely it is no longer safety but rather the essential functions of a system or a company that are managed – and managed safely – to ensure that as much as possible goes well. There is therefore something tangible to manage, be it the production of some physical artefact or the provision of some kind of service. For an airline it would be to ensure reliable, comfortable, and affordable travel; for a bank it would be to attract and serve commercial and private customers; for a hospital it would be to restore patients to health as speedily and efficiently as possible.

Managing the primary process of a system or a company well is, after all, what provides the basis for productivity and business, regardless of domain and type of activity. Therefore when something goes well – we should do our best to try to understand *why* and *how* it went well, so we can ensure that it also will go well in the future; see Figure V.3.

Safely as a social construct

Safety is itself clearly a social construct as described in Part II, not least because it is defined by its absence rather than its presence.

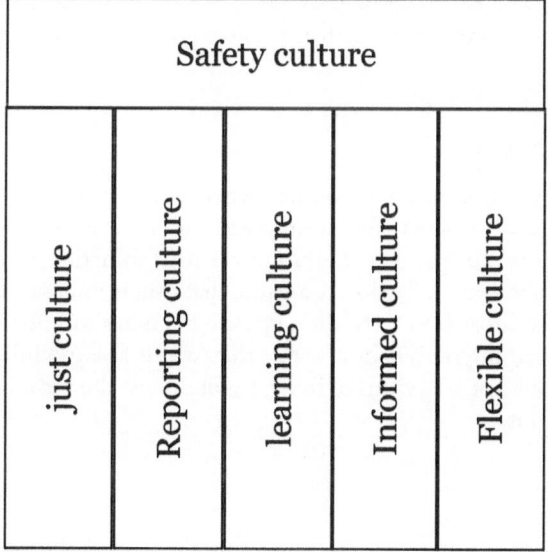

Figure V.3 A safety culture diagram (based on Reason, 1997)

Resilience and failure

A system is defined as resilient if it is able to perform as required in expected and unexpected conditions alike. (Resilience in this way is a characteristic of how a system performs, rather than a *system quality* as such. In analogy with safety the noun *resilience* should be replaced by the adverb *resilient*; Nemeth & Hollnagel, 2021). In consequence of that, failures must be exceptional events though in safety management we traditionally report only failures, hence the exceptional rather than the typical or "normal." Yet reporting might as well be guided by the need to learn rather than the other way around as discussed in Part III. Dismissing reporting as a prelude to learning – and advocating direct learning – will greatly improve both safety management and managing safely, and it is actually indispensable to the latter.

The negative stance

The negative stance represents a focus on distancing or avoiding both in learning (as avoidance learning) and in safety management. In learning it means learning what *not* to do or what to prevent, weaken, and eliminate. In safety management it means trying to ensure that as little as possible goes wrong (corresponding to vision zero).

The positive stance

The positive stance represents the focus on seeking or approaching both in learning (as reinforcement learning) and in managing safely. In learning it means learning what *to do* or strengthen, support, and facilitate. In managing safely it means trying to ensure that as much as possible goes well (corresponding to visio centum).

Safety as a modifier

The term safety is not only used on its own where it is used so frequently that no one ever asks what it actually means even though it is rarely defined but no one seems to notice or pay any attention to that (If something, such as the term safety, is meaningful to us, we simply assume that others understand, and use it in the same way we do at least until the opposite becomes too obvious to neglect. But it is also used in combination with other words as a modifier even though the combinations not always make sense. I will discuss the most common in the following sections.

Safety risk

In this often-used combination, safety is a modifier to risk. But it is not a meaningful combination as even the briefest reflection will show. Risk denotes the likelihood or probability that something unwanted or unacceptable takes place or occurs (Aven, 2006), such as the risk of an explosion or an uncontrolled release of harmful materials. It is therefore sensible if the probability or risk is as low as possible. But safety is neither an unwanted nor an unacceptable outcome, quite the opposite; the likelihood that safety is present (*occurs* is the wrong verb to use here) should therefore be as high as possible, and the usual term for that is a chance or an opportunity rather than risk, compared to the concept of visio centum that follows from a Safety-II perspective, which means that 100 percent of all activities (or as many as possible) should go well as a logical contrast to the conventional and more common vision zero that follows from a Safety-I perspective (which means that 0 percent of all activities (or as few as possible) should go wrong).

The term *safety risk* unsurprisingly also appears in ICAO's definition of safety.

Safety risk is an example par excellence of the dangers (or risks) of combining catchy terms without realising what they mean and without in any way appreciating what the combination entails. The resulting display of stupidity is in this case simply an oxymoron, which to make matters worse again is used in other equally pointless combinations such as *safety risk management*. The term really ought to be *safety possibility management* with the aim to ensure that the possibility or likelihood that the outcome of management leads to the presence of safety is as large as possible, rather than the opposite, as is it is now. No sensible organisation should be interested in safety risk management, which literally means working to ensure that the likelihood of safety is as small as possible, which is the opposite of what any sensible organisation would want. Those who nevertheless say they do safety risk management, and examples are not hard to find of that, as well as companies that have the audacity to offer courses and advice, merely demonstrate their inability to think and unintentionally demonstrate that they neither know nor care much about what they are doing.

Safety event

Another term that has come into use is *safety event*. Although it is rarely defined when it is used, it probably means an event of some sort, where safety is an issue – which paradoxically means that a safety event is most likely refer to something, a condition or an event, where safety was absent rather than present. It should therefore rightly be called a *non-safety event*. It is, for instance, used in the bow tie model (Figure III.6) as one of the initiating conditions that may contribute to a critical event.

Is risk really necessary?

This question may seem to be superfluous. Many well-respected scientists will vigorously argue that there obviously must be a relation between risk and safety. And this is correct if safety is defined as in Safety-I and vision zero: that as little as possible goes wrong, hence the absence of failures and adverse outcomes; compare this to the previous discussion of safety risk. But if safety is defined as in Safety-II and visio centum as a condition where as much as possible goes well, then risk is no longer an essential concept. It should be replaced by *chance* or *possibility*. What is of concern in managing safely is that an organisation, regardless of whether it produces refrigerators or transports people or goods between point A and point B, by land, by sea, or by air, can ensure that as much as possible goes well, hence the possibility that something goes well. Such outcomes will never be certain, due to the complexity of the activities and the multiple conditions that may affect performance as well as outcomes, as described in Part II. The outcomes can still be expressed as a probability but now for a different kind of outcome, hence a possibility rather than a risk. There is therefore a need to develop a practice of possibility assessment to complement and eventually replace risk assessment. Risk is in practice a marriage of uncertainty, unpredictability, and fear, so a divorce may be in place. But uncertainty is more than probability; it can also be a *possibility*.

Possibility theory

Possibility theory (Zadeh, 2014) is a mathematical theory for dealing with certain types of uncertainty as an alternative to probability theory. It uses measures of possibility and necessity between 0 and 1, ranging from impossible to possible and unnecessary to necessary, respectively. Professor Lotfi Zadeh first introduced possibility theory in 1978 as an extension of his theory of fuzzy sets and fuzzy logic. Dubois and Prade (1989) further contributed to its development.

Safety culture

Safety culture was first put forward as an explanation after the Bhopal disaster in 1982 and was used again as a cause or explanation when the International Safety Advisory Group (INSAG) from the International Atomic Energy Agency (IAEA) released its report on the causes of the accident at Chernobyl (INSAG-4, 1991). The two accidents, like the Three-Mile Island (TMI) accident before them, had a significant impact on the thinking about safety, because none of the generally accepted types of causes at the time (technical failures or "human error" (the latter itself a consequence of the TMI nuclear accident in 1979 and probably worse than any radiation release)) would suffice (Hale & Hovden, 1998). But human ingenuity once more came to the rescue, and a new type of cause was proposed, namely safety culture, initially defined analytically as:

> that assembly of characteristics and attitudes in organizations and individuals which establishes that, as an overriding priority (nuclear plant) safety issues receive the attention warranted by their significance.
>
> (INSAG-4, 1991, p. 5)

The purpose of Safety-I (Hollnagel, 2014b), generally defined as having no or zero accidents (Zwetsloot et al., 2013; Sharman, 2014), leads to considerable efforts being spent on determining the causes of accidents in the firm but the mistaken belief that the elimination of said causes will prevent future accidents. Safety culture also illustrates the idea of a counterfactual conditional, as described later.

Several definitions of safety culture are available and few can resist the temptation to invent a number of more specialised cultures that together constitute safety culture. The simplest and most honest definition is probably that safety culture is "the way we do things around here."

Safety management

When referring to safety management, for instance as to a safety management system (SMS), a little reflection makes clear that what is managed is not actually safety but rather how the system or company, e.g., an airline, performs so that it performs safely, meaning with as little as possible to disrupt the intended performance. It is the ability to perform safely (hence continuously) that creates revenue, rather than safety as such (safety consultants excepted, of course). So the real concern is performance management rather than safety management. The essential question is therefore not *what* is being managed (safety or the business performance of a company, but *how* it is being managed (safely, effectively, sustainably, etc., cf. Hollnagel 2020).

Safety performance

Another interesting and often-used term is *safety performance*. Performance by itself means how someone, an individual or an organisation, does something, how well they carry out a piece of work or an activity. But this is not very helpful, since safety is a state, defined either by vision zero or visio centum, rather than an activity, which is why safety grammatically is a noun.

Happily the safetyculture.com website is courageous enough to provide the following definition of the activities that comprise safety performance:

> Safety performance is the analysis of safety processes and procedures to determine how well those systems function. It involves checking the levels of risk, identifying potential hazards, evaluating safety policies and regulations, and carrying out accident investigations. It aims to reduce accidents and incidents [vision zero once more]; improve the working environment for employees; increase efficiency and productivity [which seems more like visio centum]; save time, energy, and resources by examining systems in place, diagnosing safety issues [here *safety* appears as a modifier again!], and providing solutions.

Safety performance traditionally examines various factors, including:

- Workforce experience and qualifications.
- Machine maintenance schedules.
- Safety equipment features.
- Technical specification standards.
- Compliance reporting requirements.

But just listing five factors does not really explain what safety performance is; to do so requires an account or model of how these five factors are mutually related, unless it is tacitly assumed that they are independent of each other, which hardly seems reasonable! (It is noteworthy that one of the factors, "safety equipment," itself uses *safety* as a modifier and also that *compliance* appears one more time). *Safety performance* is yet another example of a definition that makes perfect sense, provided there already is a tacit agreement of what safety means.

Three types of causes

Accident investigations in practice seem to operate happily with just three types of causes.

- The first of these is the *factual causes*, for example conditions or events that manifestly were present when an accident happened and whose existence can be established objectively after the fact. A traffic accident may for instance have been due to ice on the road or the driver being intoxicated or a tyre bursting, all three conditions that afterwards can be established with little or no uncertainty; if a factual cause had not been present, the accident would not have happened. The reasoning goes like this: "if X had not occurred, then the accident would not have happened."

- The second type is *hypothetical causes*, for example conditions that were assumed to be present but whose actual presence cannot be established after the fact. Examples are "human error" at the sharp end and organisational mistakes, which might well be considered "human error" at the blunt end. The reasoning again goes like this: "if X had not occurred, then the accident would not have happened." The hypothesis is that the accident happened because the cause was present, but the hypothesis is not one that can be verified.

- The third and most popular type of causes is *counterfactual conditionals* – or conditions – whose presence might have prevented the accident from happening. And once more the reasoning goes like this: "if only X had been present, or if only there had been more of X, or if only X had been better, then the accident would not have happened." It is counterfactual, because the accident did in fact happen so assuming X could have prevented it is contradicted by the facts. Counterfactual conditions are deceptively easy to suggest and usually also easy to understand simply because they sound plausible, which is why they are so popular and easily accepted. (Other well-known examples of counterfactual conditions are situation awareness, communication, leadership, and trust.) Although counterfactual conditions do require an articulated theoretical basis, they rarely have one from the beginning. Counterfactual conditions also illustrate the efficiency-thoroughness trade-off principle (Hollnagel, 2009), they are efficient and require little or no effort (intellectual or otherwise to create and disseminate), but they totally lack thoroughness since they just name a cause but do not explain anything at all. It is hardly surprising that, whenever a new concept is proposed, attempts are made to explain and understand it by decomposing it, by breaking it into parts. Decomposition is a time-honoured and nearly irresistible tradition of Western science that has been used successfully for the physical world, since it was first proposed by the ancient Greek philosopher Democritus (c. 460–370 BCE). Democritus, a student of Leucippus (5th century BCE) formulated the atomic theory of the universe, according to which any particle, no matter how small, can always be divided into smaller particles until we reach the smallest indivisible part named the atom. Even though it is less certain that the principle also works for the non-physical world, for concepts and ideas, that has never stopped people from trying to do so, and safety culture is no exception, for instance Gilbert et al. (2016), among many others. In the physical world the decomposition stops at the smallest component, either the invisible or the indivisible. In the non-physical world the decomposition stops when the thought reaches the unthinkable, although human ingenuity usually manages to go well beyond that by inventing new ideas (Cooper, 2018).

One of the better-known examples of such decomposition is James Reason's (1997) proposal that safety culture contains four components or subcultures (Figure V.3): The components are called: a reporting culture, a just culture, a flexible culture, and a learning culture, respectively. The obvious question is: Which of these can be considered a safety culture atom, in the sense that it cannot be decomposed further?

In addition to being a counterfactual condition, safety culture is also a good example of a monolithic cause, as described in Part II. In the case of safety culture there is, unfortunately, no known silver bullet as an effective monolithic antidote or solution.

Alternative approaches

Mindfulness

A different way to capture the contribution of culture to organisational performance has been to define certain core values corresponding to Schein's concept of espoused values (Schein, 1992) which should be shared; foremost among these is mindfulness (Weick & Sutcliffe, 2001), meaning the readiness continuously to scrutinise existing and emerging expectations within a larger context. A mindful culture or, to use Reason's (1997) terms, an informed culture, contains the four components shown in Figure V.3. A mindful culture can therefore not be the sought-after safety atom. The latter two subcultures refer to an organisation's adaptive capabilities, supporting, for instance, good working practices that allow teams to recognise changing environmental conditions and switch modes of operation accordingly.

Just culture

Since practically all efforts to improve safety focus on the prevention of accidents, information is obviously needed about why and how accidents happen. Here accident investigations are of limited value, since they usually focus on the factual events in physical terms. It is therefore also necessary to know what people did before and during the accident, paradoxically called *safety behaviours*. Such information is, however, hypothetical rather than factual; even with current smart watches and devices, we surely need the cognoscope that Crovitz (1970) imagined – but who knows what the future may bring? At present, the only available sources of such information are the people who were on the spot, at the coalface, in the cockpit, or in the operating room. But since it for obvious reasons can be difficult to get people to tell about something they have done when they know it may be used against them, then they need to be protected against possible blame. There is consequently a derived need for some sort of assurance that makes it possible to divulge such information. This leads to the inevitable conclusion that there is a need for a just culture, which clearly is a counterfactual conditional, if ever there was one. The UK's National Air Traffic Services (NATS) in 2004 issued a strategic plan for safety which mentions just culture more than 100 times without once offering a definition of what it is, except indirectly in the following passage:

> safety performance [sic!] is measured through a comprehensive incident reporting and investigation process. Safety incidents [sic!]. . . are reported through the Mandatory Occurrence Report scheme used by the Safety Regulation Group (SRG) of the Civil Aviation Authority (CAA), and investigated and assessed by both NATS and, if necessary, by the SRG.
>
> NATS is committed to maintaining a "just" reporting culture to ensure that all safety related incidents [sic!] continue to be reported and investigated.

While this provides the justification for just culture it is by no means an articulated premise. The last step introduces just culture as a counterfactual conditional for safety culture.

The arguments presented by NATS hide the important assumption that safety depends on people reporting what goes wrong, learning from accidents, even though one might as well and more easily learn from what has gone well. Yet the problems in getting information from people magically disappears once it is realised that the performance of interest actually is what Weick (1987) called dynamic non-events, defined as the "conditions in which problems are momentarily under control due to compensating changes." It is, actually quite easy to get people to talk about their work. (Hollnagel, Shorrock & Johns, 2021) provide several examples of how this has been done successfully. Managing safely must refer to the everyday operations that we usually fail to pay attention to, precisely because they happen all the time without being spectacular in any way except by their sheer number (or frequency). Anecdotal evidence (from health care and aviation) suggests that the problem is less to get people to talk than to make them stop once they have started.

What should managing safely be based on?

Part I of this book (Legacy) characterised and analysed the basis for safety management and found it wanting. The obvious next question is: What should the basis be for managing safely? It is hardly surprising that the legacy from the 1930s needs to be revised and updated. The world today is non-trivial rather than trivial, and methods and models must obviously acknowledge that. The object of management in managing safely is the main activities of the system or company in question. Part IV argued that it is necessary to know what the current state or position is, what the goal or target is, and also what the effective means to make a change are. In order to manage an activity or process it is necessary to acknowledge the difference between WAI – WAD, to abandon linearity and replace causality by emergence (as both George Henry Lewes and Bertrand Russell suggested), to abandon causality and linear flow models, to recognise the human as an asset rather than a liability – which is hardly news (Le Coze, 2022), and also to acknowledge that safety cannot be isolated – hence to use the adverb *safely* rather than the noun *safety* – and overall to avoid the simplifications in thinking and doing that is an inseparable part of the safety legacy.

Requisite imagination

A theme in the previous parts that will appear again in several other contexts is the importance of imagination. Adamski and Westrum (2003) introduced the concept of requisite imagination in analogy with the concept of requisite variety known from cybernetics. Requisite variety is the variety a regulator must have to match the variety of the processes being regulated. Requisite imagination is the imagination needed to be able to consider all that can happen in practice.

The fine art of anticipating what might go wrong means taking sufficient time to reflect on the design in order to identify and acknowledge potential problems. We call this fine art *requisite imagination* (Adamski & Westrum, 2003, p. 4). Given the time when this was proposed (in 2004), it is hardly surprising that the reference was to problems and to what could go wrong. Today, nearly 20 years later, one might define *requisite imagination* as the ability to anticipate how something may happen or be implemented, which requires that sufficient time is taken to reflect on the design in order to identify and acknowledge potential issues and dependencies of concern. To be fair, Adamski and Westrum also offered the following more neutral formulation: "requisite imagination is the ability to imagine key aspects of the future we are planning" (Westrum, 2006).

Events and non-events

It was Karl Weick who in 1987 suggested that the term *dynamic non-event* be used to describe work that goes well, defined as "an ongoing condition in which problems are momentarily under control due to compensating changes." This also reflects the common attitude that, if there have been no accidents during a period, then "nothing" has happened, or as Reason (2000) noted, "Safety is defined and measured more by its absence than by its presence," which actually means that whenever something has gone wrong people are prone to say that there was an absence of safety – or that safety was lacking (The underlying "logic" is that if safety had been present then the accident would not have happened, but this makes safety a counterfactual condition). We notice when something goes wrong, because it is unexpected, and such events are strong signals. But we rarely notice when things go well, because it is what we expect, and such daily events are weak signals. The reason why we do not notice them is because humans have a strong tendency to habituate to whatever happens all the time, such as background noises or smells and unfortunately also to work that goes well, because that is actually what happens most of the time.

Habituation

Habituation is the decline in responsiveness to a stimulus due to repeated or constant exposure. In physiology it is period of time, the *refractory period*, during which an organ or cell is incapable of repeating a particular action. Habituation explains how we habituate to repeated sensory stimulation. A simple example is if you have to stay in a place where the ambient noises are different from what you are used to, for instance at a beach hotel or in a (small Dutch) town with many churches, hence bells ringing incessantly or just different traffic noises – or even the absence of noises (aka silence). Other examples are different smells, as when you enter a hospital or a burger restaurant. In the beginning you notice the noises and/or smells, but you soon get used to them and then do not notice them any longer. But there also seems to be a cognitive refractory period, which is more like semantic satiation or reactive inhibition. Habituation is a very useful psychological "mechanism" or response from the nervous system, since it is only evolutionarily useful to respond to what is new or different, (cf., the concept of the jnd) but habituation unfortunately works for all kinds of repeated events or stimuli, including work that goes well. Another example is that when you sit down on a chair or some other support you are acutely aware to begin with of the pressure of the support against your body, but after a short while you do not notice it at all.

Walking through a crowd

A good example dynamic non-events is walking through a crowd of people, for instance in a shopping centre, a railway station, an airport, or a crowded sidewalk. It is something we all have done hundreds if not thousands of times. The rare event is when we bump into someone else, or someone else bumps into us. But we typically just walk through the crowd to get to an exit or another specific location. Normally this is uneventful, and if asked what happened, most people would doubtless say that nothing happened, meaning that they did not collide with someone else. But even a brief reflection makes clear that many things happened; we, as well as the people around us, changed speed, gait, and direction an almost countless number of times. We hurried through a sudden opening, like a field of safe passage (Gibson & Crooks, 1938), we slowed down or stopped to let someone pass, but we did it all so smoothly and automatically that no one noticed it. All the tiny adjustments we made were truly non-events, but it was only because of these that we and others successfully reached our target, that we did not bump into someone else. It is very instructive to observe this and pay attention to the multitude of non-events, not as you walk through the crowd, but it may be wise to stand aside while you do so. It is also instructive to notice the difference between those who have not yet acquired the skill, like children and those who are skilled, such as most adults. It may, of course, have become less smooth since the appearance of the smartphone, which means that many people now walk without paying attention to what happens around them, almost like children. One possible reason why we do not notice it is that it is a very funda-mental skill that most animals also possess; think for instance of a stampeding herd of cattle – at least as we have seen it in the movies – or cows being let out into the fields after having spent the winter in the stables. (The animals do admittedly generally run in the same direction, rather than crisscrossing each other.)

五輪書 *(Go rin no sho, The book of five rings)*

Another reference to the importance of non-events is found in *The book of five rings*, a famous text on swordsmanship and the martial arts in general, written by the legendary Japanese swordsman Miyamoto Musashi around 1645. (The book is freely available from several sites on the internet.)

Dueling with a long samurai sword can hardly be considered a safe activity; in fact about half of those who ever tried it died, at least until it became commonly accepted that a duellist could yield (but then there was still the likelihood of suf-fering some harm). Yet Musashi is known to have fought and won 61 duels. So he clearly was an expert in personal safety, and it may therefore be a good idea to pay attention to his advice, which fortunately can be found in *The book of five rings*. His primary advice was

- Perceive those things which cannot be seen. (The things that cannot be or which are not seen correspond to the dynamic non-events that Weick wrote about 342 years later).

Musashi also made a distinction between looking and perceiving, where the lat-ter implies understanding and giving meaning to what you are looking at.

A second piece of advice was

- Pay attention even to trifles.

Coda epilogue

In the way we manage safety as well as other aspects of organisational perfor-mance we tend to rely on rules of thumb – not least because doing so makes life easier. And here is another rule of thumb, do not rely on rules of thumb but heed the advice of Lord Mountbatten of Burma: "It is no good getting the right answer to the wrong questions: you've got to get the right question before the right answer can be of any use." So what are the right questions for safety management and managing safely? Are they even the same? No, they are not. The right question for safety management is represented by vision zero. And the right question for managing safely is represented by visio centum. And being different questions they must, of course, have different answers, but which question is the right one? Which question does best enhance the ability of a system or a company to function as required in expected and unexpected conditions alike (in other words to be resilient)? The easiest way in the short run may not be the best solution in the long run, and safety management and managing safely should both be activities that are for the long run. Another version of that is that since we generally find only what we are looking for and also generally only fix the problems that we find, it is important to expend some effort to look for the right problems from the start, rather than rely on the canned "wisdom" of the safety legacy.

Acknowledgments

The idea of the four potentials first saw the light of day as the four cornerstones of resilience engineering (Hollnagel, 2009) that were developed into a method called the Resilience Analysis Grid (RAG) (Hollnagel, 2017). This became the four potentials during the writing of Hollnagel et al. (2021). The contributions from Toni and Jörg, as well as the support of Eurocontol, are gratefully acknowledged, including the permission to reuse the material as the basis for this little book. During the planning and writing I also received encouragement and comments from Professor Sidney Dekker and Professor Andrew Sharman. I apologise If I have not always followed their advice. And finally, I would like to acknowledge the many colleagues with whom I have argued and often disagreed over the years. Their names are too many for my dwindling memory and far too many to mention here. It would read like a who's who in safety (in some cases unfortunately a who-was-who) but the main events have been the EHS meetings held in Berlin and the Safety-II in practice meetings held at various places on the globe, as well as the FRAMily meetings and the annual conferences of the resilient health care society. May all continue for many years to come.

I would also like to declare a long overdue debt of gratitude to David Klahr who many years ago tried to teach me how effectively to organise a written argument.

General epilogue

The only way to become better at what you do is to pay attention to what actually happens when it goes well and use that to improve how it was done: Try to investigate and understand what made acceptable performance possible. In other words, acknowledge that thoroughness is a precondition for efficiency.

And remember that while simple problems may have correspondingly simple solutions, complex problems practically always require correspondingly complex solutions. Safety management is often presented as a simple solution. Managing safely is definitely not a simple solution: Disguising complex problems as simple problems by offering apparently simple solutions does not make the problems any simpler but will more likely guarantee that the solutions will not be effective.

References

The following list of references only includes works that I refer to directly in the book. In some cases quoted works themselves refer to other works, such indirect or secondhand references are not included.

Adamski, A. J., & Westrum, R. (2003). Requisite imagination: The fine art of anticipating what might go wrong. In E. Hollnagel (Ed.), *Handbook of cognitive task design*. Hillsdale, NJ: Lawrence Erlbaum Associates.

Allen, J. F. (1983). Maintaining knowledge about temporal intervals. *Communications of the ACM, 26*(11), 832–843.

Amalberti, R. (2001). The paradoxes of almost safe transportation systems. *Safety Science, 37,* 109–126.

American Heritage. (2008). *Stedman's medical dictionary*. Boston, MA: Houghton Mifflin Company.

American Society of Safety Engineers (ASSE). (2011). *Prevention through design guidelines for addressing occupational hazards and risks in design and redesign process* (ANSI/ASSE Z590.3). ASSE. https://webstore.ansi.org/standards/asse/ansiassez5902011

Annett, J. (1972). *Feedback and human behaviour*. Harmondsworth: Penguin Education/Penguin Books.

Ansoff, H. I. (1975). Managing strategic surprise by response to weak signals. *California Management Review, 18*(2) 21–33.

Argyris, C. (1977). Double loop learning in organizations. *Harvard Business Review, 55*(5), 115–125.

Ashby, W. R. (1956). *An introduction to cybernetics*. London: Chapman & Hall, Ltd.

Aven, T. (2006). A unified framework for risk and vulnerability analysis covering both safety and security. *Reliability Engineering & System Safety, 92*(6), 745–754.

Aven, T. (2022). A risk science perspective on the discussion concerning Safety I, Safety II and Safety III. *Reliability Engineering & System Safety, 217*, 108077.

BBC. (2018, March 14). *United Airlines responsible for dog death in overhead locker*. https://www.bbc.com/news/world-us-canada-43394952

Beer, S. (1959). *Cybernetics and management*. London: The English Universities Press.

Beer, S. (1966). *Decision & control. The meaning of operational research and management cybernetics*. London: John Wiley & Sons.

Beer, S. (1984). The viable system model: Its provenance, development, methodology and pathology. *Journal of the Operational Research Society, 35*(1), 7–25.

Bloch, A. (2003). *Murphy's law*. Harmondsworth: Penguin Books.

Bohn, R. (2000). Stop fighting the fires. *Harvard Business Review, 78*(4), 83–92.

Bradshaw, J. M., Hoffman, R. R., Woods, D. D., & Johnson, M. (2013). The seven deadly myths of "autonomous systems". *IEEE Intelligent Systems, 28*(3), 54–61.

Brehmer, B. (2007). Understanding the functions of C2 is the key to progress. *The International C2 Journal, 1*(1), 211–232.

Buckley, W. (Ed.). (1968). *Modern systems research for the behavioral scientist.* Chicago, IL: Aldine.

Cilliers, P. (2005). Complexity, deconstruction and relativism. *Theory, Culture & Society, 22*(5), 255–267.

Cilliers, P., & Richardson, K. (2001). Special editors' introduction: What is complexity science? A view from different directions. *Emergence: Complexity and Organization*, 3(1), 5-24.

Cojazzi, G., & Pinola, L. (1994, March 20–25). Root cause analysis methodologies: Trends and needs. In G. E. Apostolakis & J. S. Wu (Eds.), *Proceedings of PSAM-II.* Amsterdam: Elsevier Science.

Conant, R. C., & Ashby, W. R. (1970). Every good regulator of a system must be a model of that system. *International Journal of Systems Science, 1*(2), 89–97.

Conklin, T. (2012). *Pre-accident investigations. An introduction to organizational safety.* Boca Raton, FL: CRC Press.

Coombs, C. H., Dawes, R. M., & Tversky, A. (1970). *Mathematical psychology.* Englewood Cliffs, NJ: Prentice Hall, Inc.

Cooper, M. D. (2018). The safety culture construct: Theory and practice. In C. Gilbert, B. Journé, H. Laroche, & C. Bieder (Eds.), *Safety cultures, safety models: Taking stock and moving forward* (pp. 47–61). Cham: Springer Nature.

Cooper, M. D. (2022). The emperor has no clothes: A critique of Safety-II. *Safety Science, 152.* https://doi.org/10.1016/j.ssci.2020.105047

Cowan, N. (2008). What are the differences between long-term, short-term, and working memory? *Progress in Brain Research, 169,* 323–338.

Crovitz, H. E. (1970). *Galton's walk.* New York: Harper & Row.

Dekker, S. (2015). *Safety differently. Human factors for a new era.* Boca Raton, FL: CRC Press.

de La Mettrie, J. O. (1996, org. 1747). *Machine man and other writings.* Cambridge: Cambridge University Press.

de Laplace, P. S. (1820). *Theorie Analytique des probabilités. Troisième édition, revue et augmenteè par l'ateur.* Paris: Courcier.

Deming, W. E. (1950). *Elementary principles of the statistical control of quality: A series of lectures.* Tokyo: Union of Japanese Scientists and Engineers.

de Ruijter, A., & Guldenmund, F. (2016). The bowtie method: A review. *Safety Science, 88,* 211–218.

Dougherty, E. M. Jr. (1990). Human reliability analysis – where shouldst thou turn? *Reliability Engineering and System Safety, 29*(3), 283–299.

Dubois, D., & Prade, H. (1989). Handling uncertainty in expert systems – pitfalls, difficulties, remedies. In E. Hollnagel (Ed.), *The reliability of expert systems* (pp. 64–99). Chichester: Ellis Horwood Ltd.

Engeström, Y., Miettinen, R., & Punamäki, R.-L. (1999). *Perspectives on activity theory.* Cambridge: Cambridge University Press.

Fischhoff, B. (1975). Hindsight ≠ foresight: The effect of outcome knowledge on judgment under uncertainty. *Journal of Experimental Psychology, Human Perception and Performance, 1*(3), 288–299.

Fitts, P. M., Chapanis, A., Frick, F. C., Garner, W. R., Gebhard, J. W., Henneman, R. H., Koppauf, W. E., Newman, E. B., & Williams, A. C. Jr. (Eds.). (1951). *Human engineering for an effective air navigation and traffic-control system.* Columbus, OH: Ohio State University Research Foundation.

Flin, R. (2006). Erosion of managerial resilience: From Vasa to NASA. In E. Hollnagel, D. D. Woods, & N. Leveson (Eds.), *Resilience engineering: Concepts and precepts* (pp. 223–233). Boca Raton, FL: CRC Press.

Gehman, H. W. (2003). *Columbia Accident Investigation Board: (Issued with CD-ROM)* (Vol. 2). Washington, DC: National Aeronautics and Space Administration and the Government Printing Office.

Gibson, J. J., & Crooks, L. E. (1938). A theoretical field-analysis of automobile-driving. *The American Journal of Psychology, 51*(3), 453–471.

Gilbert, C., Journé, B., Laroche, H., & Bieder, C. (2016). *Safety cultures, safety models: Taking stock and moving forward* (p. 166). Cham: Springer Nature.

Hale, A. (2002). Conditions of occurrence of major and minor accidents Urban myths, deviations and accident scenarios. *Tijdschrift voor toegepaste Arbowetenschap, 15*(3), 34–40.

Hale, A., Goossens, L., Ale, B., Bellamy, L., Post, J., Oh, J., & Papazoglou, I. A. (2004). Managing safety barriers and controls at the workplace. In *Probabilistic safety assessment and management: PSAM 7 – ESREL'04 June 14–18, 2004, Berlin, Germany* (Vol. 6, pp. 608–613). London: Springer.

Hale, A. R., & Glendon, A. I. (1987). *Individual behaviour in the control of danger*. London: Elsevier Science.

Hale, A. R., & Hovden, J. (1998). Management and culture: The third age of safety. A review of approaches to organisational aspects of safety, health and environment. *Occupational Injury*, 145–182.

Heinrich, H. W. (1931). *Industrial accident prevention*. New York: McGraw-Hill Insurance Series.

Heinrich, H. W. (1959). *Industrial accident prevention. A scientific approach* (4th ed.). New York: McGraw-Hill Book Company, Inc.

Heinrich, H. W., Peterson, D., & Roos, N. (1980). *Industrial accident prevention: A safety management approach* (5th ed.). New York: McGraw Hill.

Hobbes, T. (1651). *Leviathan or the matter, forme, & power of a common-wealth ecclesiasticall and civill*. London: Andrew Croe.

Hollnagel, E. (1993). The phenotype of erroneous actions. *International Journal of Man-Machine Studies, 39*(1), 1–32.

Hollnagel, E. (1998). *Cognitive reliability and error analysis method (CREAM)*. London: Elsevier Science.

Hollnagel, E. (2002). Understanding accidents-from root causes to performance variability. In *Proceedings of the IEEE 7th conference on human factors and power plants* (pp. 1–6). NJ: IEEE.

Hollnagel, E. (2004). *Barriers and accident prevention*. London: Routledge.

Hollnagel, E. (Ed.). (2004a). *Handbook of cognitive task design* (pp. 193–220). Boca Raton, FL: CRC Press.

Hollnagel, E. (2009). *The ETTO principle: Why things that go right sometimes go wrong*. Boca Raton, FL: CRC Press.

Hollnagel, E. (2012). *FRAM: The functional resonance analysis method: Modelling complex socio-technical systems*. Boca Raton, FL: CRC Press.

Hollnagel, E. (2014a). Is safety a subject for science? *Safety Science, 67*, 21–24.

Hollnagel, E. (2014b). *Safety-I and Safety-II: The past and future of safety management*. Boca Raton, FL: CRC Press.

Hollnagel, E. (2017). *Safety-II in practice*. London: Routledge.

Hollnagel, E. (2020). *Synesis: The unification of productivity, quality, safety and reliability*. London: Routledge

Hollnagel, E., Licu, A., & Leonhardt, J. (2021). *The systemic potentials management: Building a basis for resilient performance* (A White paper). Brussels: Eurocontrol. https://skybrary.aero/bookshelf/systemic-potentials-management-building-basis-resilient-performance.

Hollnagel, E., Shorrock, S., & Johns, A. (2021). *Learning from all operations: Expanding the field of vision to improve aviation safety* (White paper). Alexandria, VA: A Flight Safety Foundation.

Hollnagel, E., & Woods, D. D. (1983). Cognitive systems engineering: New wine in new bottles. *International Journal of Man-Machine Studies, 18*(6), 583–600.

Hollnagel, E., & Woods, D. D. (2005). *Joint cognitive systems: Foundations of cognitive systems engineering.* Boca Raton, FL: CRC Press.

Hudson, P.T., Parker, D., Lawton, R., & van der Graaf, G. C. (2008). *Managing non-compliance: Moving from theory to practice.* Paper presented at the SPE international conference and exhibition on health, safety, environment, and sustainability? (Paper Number: SPE-73992). SPE.

Hudson, P.T., Parker, D., & van der Graaf, G. C. (2002). *The hearts and minds program: Understanding HSE culture.* Paper presented at the SPE International Conference on Health, Safety and Environment in Oil and Gas Exploration and Production (Paper Number: SPE-73938). SPE.

Hume, D. (1739). *A treatise of human nature.* London: John Noon.

INSAG-4. (1991). *Safety culture – a report by the International Nuclear Safety Advisory group.* Vienna: International Atomic Energy Agency.

International Civil Aviation Organization (ICAO). (2013). *Safety Management Manual (SMM)* (Doc 9859 AN/474). Montreal: ICAO.

James, W. (1890). *The principles of psychology.* London: Macmillan.

Jaubert, J.-M. (2006). The meaning of safety. *Offshore Technology.* https://www.offshore-technology.com/features/feature577/.

Kanigel, R. (2005). *The one best way: Frederick Winslow Taylor and the enigma of efficiency.* Boston, MA: The MIT Press.

Klein, G. (2011). Critical thoughts about critical thinking. *Theoretical Issues in Ergonomics Science, 12*(3), 210–224.

Klein, G. A. (1993). A recognition-primed decision (RPD) model of rapid decision making. In G. A. Klein, J. Orasanu, R. Calderwood, & C. E. Zsambok (Eds.), *Decision making in action: Models and methods* (pp. 138–147). Norwood, NJ: Ablex.

Klein, G. A. (1998). *Sources of power: How people make decisions.* Cambridge, MA: MIT Press.

Kletz, T. A. (2001). *Learning from accidents* (3rd ed.). London: Routledge.

Landsman, K., & Van Wolde, E. (2016). *The challenge of chance: A multidisciplinary approach from science and the humanities* (p. 276). Cham: Springer Nature.

Le Coze, J.-C. (2022). The 'new view' of human error. Origins, ambiguities, successes and critiques. *Safety Science, 154.* https://doi.org/10.1016/j.ssci.2022.105853.

Lehto, M., & Salvendy, G. (1991). Models of accident causation and their application: Review and reappraisal. *Journal of Engineering and Technology Management, 8*(2), 173–205.

Leveson, N. (2016). http:/psas.scripts.mit.edu/home/wp-content/uploads/2016/04/STAMP-Intro-2016.pdf (accessed April 13, 2024).

Leveson, N. (2020). *Safety III: A systems approach to safety and resilience.* MIT Engineering Systems Lab. Sunnyday.mit.edu/safety-3.Pdf (accessed April 13, 2024).

Leveson, N. G. (1992). High-pressure steam engines and computer software. In *Proceedings of the 14th international conference on software engineering* (pp. 2–14). New York: The Association for Computing Machinery.

Leveson, N. G. (1995). *Safeware: System safety and computers.* New York: Association for Computing Machinery.

Lewin, K. (1952). Group decision and social change. In E. Newcombe & R. Harley (Eds.), *Readings in social psychology* (pp. 459–473). New York: Henry Holt.

Lind, M. (2003). Making sense of the abstraction hierarchy in the power plant domain. *Cognition, Technology & Work, 5,* 67–81.

Lindblom, C. E. (1959). The science of "muddling through". *Public Administration Review, 19,* 79–88.

Lindsay, P. H., & Norman, D. A. (1972). *Human information processing, An introduction to psychology.* New York: Academic Press.

Lundberg, J., Rollenhagen, C., & Hollnagel, E. (2009). What-you-lo-for-is-what-you-find – the consequences of underlying accident models in eight accident investigation manuals. *Safety Science, 47,* 1297–1311.

Mach, E. (1908). *Erkenntnis und Irrtum – Skizzen zur Psychologie der Forschung.* Leipzig: Verlag von Johann Ambrosius Bach.

MacKay, D. M. (1956). Towards an information-flow model of human behaviour. *British Journal of Psychology, 47*(1), 30–43.

March, F. G., & Simon, H. A. (1958). *Organizations.* New York: John Wiley.

Marshall, E. C., Duncan, K. D., & Baker, S. M. (1981). The role of withheld information in the training of process plant fault diagnosis. *Ergonomics, 24*(9), 711–724.

Maruyama, M. (1963). The second cybernetics: Deviation-amplifying mutual causal processes. *American Scientist,* 164–179.

Merton, R. K. (1936). The unanticipated consequences of purposive social action. *American Sociological Review, 1*(6), 894–904.

Merton, R. K., & Barber, E. (2011). *The travels and adventures of serendipity: A study in sociological semantics and the sociology of science.* Princeton, NJ: Princeton University Press.

Miller, G. A. (1956). The magical number seven, plus or minus two: Some limits on our capacity for processing information. *Psychological Review, 63*(2), 81.

Miller, G. A., Galanter, E., & Pribram, K. H. (1960). *Plans and the structure of behaviour.* New York: Freeman.

Miller, J. G. (1960). Information input overload and psychopathology. *American Journal of Psychiatry, 116*(8), 695–704.

Moen, R. D., & Norman, C. L. (2010). Circling back. *Quality Progress, 43*(11), 22.

Moray, N. P. (1967). Where is capacity limited? A survey and a model. *Acta Psychologica, 27,* 84–92.

Morozow, E. (2013a). The perils of perfection. *The New York Times,* 2 March.

Morozow, E. (2013b). *To save everything, click here: The folly of technological solutionism.* New York: The Perseus Books Group.

Morris, M. W., Moore, P. C., & Sim, D. L. (1998). *Counterfactual thinking about accidents and the "human error" fallacy: How "undoing" accidents leads decision makers to futile human-focused remedies.* Graduate School of Business, Stanford University.

Mosby's Medical Dictionary. (2009). (8th ed.). Elsevier.

National Air Traffic Services (NATS). (2004). Safety plan 2017–2019. Advancing aviation keeping the skies safe. https://www.nats.aero/wp-content/uploads/2012/07/Safety-Plan-2017-19.pdf

Neisser, U. (1967). *Cognitive psychology.* New York: Appleton-Century.

Nemeth, C. P., & Hollnagel, E. (Eds.). (2021). *Advancing resilient performance.* Cham: Springer Nature.

Newell, A. (1990). *Unified theory of cognition.* Cambridge, MA: Harvard University Press.

Newell, A., & Simon, H. A. (1963). GPS, A program that simulates human thought. In E. A. Feigenbaum & J. Feldman (Eds.), *Computers and thought* (pp. 279–293). New York: McGraw-Hill.

Newell, A., & Simon, H. A. (1972). *Human problem solving*. Englewood Cliffs, NJ: Prentice-Hall.

Nietzsche, F. (1977, org. 1878). *Twilight of the idols. Or, how to philosophize with the hammer*. Indianapolis, IN; Cambridge: Hackett Publishing Company, Inc.

Pastore, R. E., & Scheirer, C. J. (1974). Signal detection theory: Considerations for general application. *Psychological Bulletin, 81*(12), 945.

Patriarca, R., Di Gravio, G., Woltjer, R., Costantino, F., Praetorius, G., Ferreira, P., and Hollnagel, E. (2017). Framing the FRAM: A literature review on the functional resonance analysis method. *Safety Science, 91,* 49–60.

Paxton, L. J. (2007). "Faster, better, and cheaper" at NASA: Lessons learned in managing and accepting risk. *Acta Astronautica, 61*(10), 954–963.

Peirce, C. S. (2014, org. 1878). *Illustrations of the logic of science: How to make our ideas clear* (Edited by Cornelis de Waal). Chicago, IL: Open Court.

Perrow, C. (1984). *Normal accidents: Living with high-risk technologies.* Princeton, NJ: Princeton University Press.

Peterson, W. W., Birdsall, T., & Fox, W. C. (1954). The theory of signal detectability. *Transactions of the IRE Professional Group on Information Theory, 4*(4), 171–212. doi:10.1109/TIT.1954.1057460

Popper, K. (1959). *The logic of scientific discovery.* London: Hutchinson and Co.

Pringle, J. W. S. (1951). On the parallel between learning and evolution. *Behaviour, 3,* 174–215.

Rapoport, A. (1970). *N-person game theory: Concepts and applications.* Ann Arbor, MI: The University of Michigan Press.

Rasmussen, J. (1974). *The human data processor as a system component. Bits and pieces of a model* (Risø-M-1722). Roskilde: Risø National Laboratory.

Rasmussen, J. (1983). Skills, rules, and knowledge; signals, signs, and symbols, and other distinctions in human performance models. *IEEE Transactions on Systems, Man, and Cybernetics,* (3), 257–266.

Rasmussen, J. (1987). *Information processing and human-machine interaction. An approach to cognitive engineering.* North-Holland.

Rasmussen, J., Duncan, K., & Leplat, J. (Eds.). (1987). *New technology and human error.* New York: Wiley.

Rasmussen, J., & Lind, M. (1981). *Coping with complexity* (Risø-M-No. 2293). Roskilde: Risø National Laboratory.

Reason, J., Hollnagel, E., & Paries, J. (2006). Revisiting the Swiss cheese model of accidents. *Journal of Clinical Engineering, 27*(4), 110–115.

Reason, J. T. (1990a). *Human error.* Cambridge: Cambridge University Press.

Reason, J. T. (1990b). The contribution of latent human failures to the breakdown of complex systems. *Philosophical Transactions of the Royal Society of London B: Biological Sciences, 327*(1241), 475–484.

Reason, J. T. (1997). *Managing the risks of organizational accidents.* Aldershot: Ashgate Publishing Limited.

Reason, J. T. (2000). Safety paradoxes and safety culture. *Injury Control & Safety Promotion, 7*(1), 3–14.

Rouse, W. B., & Rouse, S. H. (1979). Measures of complexity of fault diagnosis tasks. *IEEE Transactions on Systems, Man, and Cybernetics, 9*(11).

Russell, B. (1913). On the notion of cause. *Proceedings of the Aristotelian Society, 13,* 1–26.

Santally, M. I., Cooshna-Naik, D., Conruyt, N., & Wing, C. K. (2015). A social partnership model to promote educators' development in Mauritius through formal and informal capacity-building initiatives. *Journal of Learning for Development, 2*(1).

Schein, E. (1992). *Organizational culture and leadership.* San Francisco, CA: Jossey Bass.

Schoemaker, P. J., & Day, G. S. (2009). How to make sense of weak signals. Leading Organizations: Perspectives for a new era, 37. *MIT Sloan Management Review, 50*(3).

Searle, J. R. (1995). *The construction of social reality.* New York: Simon and Schuster.

Senders, J. W., & Moray, N. P. (1991). *Human error. Cause, prediction, and reduction.* Hillsdale, NJ: Lawrence Erlbaum.

Shappell, S. A., & Wiegmann, D. A. (2000). *The human factors analysis and classification system—HFACS.* Daytona Beach, FL: Embry Riddle Aeronautical University.

Sharman, A. (2014). *From accidents to zero: A practical guide to improving your workplace safety culture.* Paris: Maverick Eagle Press.

Shewhart, W. A. (1931). *The economic control of quality of manufactured product.* New York: D. Van Nostrand Company.

Simard, S. (1921). *Finding the mother tree: Discovering the wisdom of the forest.* New York: Knopf.

Simon, H. A. (1947). *Administrative behavior: A study of decision-making processes in administrative organization* (1st ed.). New York: Macmillan.

Simons, D. J., & Rensink, R. A. (2005). Change blindness: Past, present, and future. *Trends in Cognitive Sciences, 9*(1), 16–20.

Taylor, F. W. (1911). *The principles of scientific management.* New York: Harper.

Taylor, J. R. (1976). *Interlock design using fault tree and cause consequence analysis* (Risø-M-1890). Roskilde: Risø national Laboratory.

Tiessen, B. (2008). On the journey to a culture of patient safety. *Healthcare Quarterly, 11*(4), 58–63.

Tolstoy, L. (2007, org. 1869). *War and peace.* London: Penguin Classics.

Tversky, A. (1972). Elimination by aspects: A theory of choice. *Psychological Review, 79*(4), 463–492.

Tversky, A., & Kahneman, D. (1974). Judgment under uncertainty: Heuristics and biases. *Science, 185*(4157), 1124–1131.

Urry, J. (2005). The complexity turn. *Theory, Culture & Society, 22*(5), 1–14.

Vicente, K. J., & Rasmussen, J. (1988). On applying the skills, rules, knowledge framework to interface design. *Proceedings of the Human Factors Society Annual Meeting, 32*(5), 254–258.

Von Neumann, J., & Morgenstern, O. (1944). *Theory of games and economic behavior.* Princeton, NJ: Princeton University Press.

Wagoner, B. (2013). Bartlett's concept of schema in reconstruction. *Theory & Psychology, 23*(5), 553–575.

Weick, K. E. (1987). Organisational culture as a source of high reliability. *California Management Review, 29*(2), 112–128.

Weick, K. E., & Sutcliffe, K. M. (2001). *Managing the unexpected.* San Francisco, CA: Jossey-Bass.

Weinberg, G. M., & Weinberg, D. (1979). *On the design of stable systems.* New York: Wiley.

Westrum, R. (2006). A typology of resilience situations. In E. Hollnagel, D. D. Woods, & N. Leveson (Eds.), *Resilience engineering: Concepts and precepts* (pp. 55–65). Boca Raton, FL: CRC Press.

Wickens, C. (1992). *Engineering psychology and human performance* (2nd ed.). New York: Harper-Collins.

Wienen, H. C. A., Bukhsh, F. A., Vriezekolk, E., & Wieringa, R. J. (2017). *Accident analysis methods and models-a systematic literature review.* Centre for Telematics and Information Technology.

Wiener, N. (1964). *God & Golem, Inc.: A comment on certain points where cybernetics impinges on religion.* Cambridge, MA: MIT Press..

Wittgenstein, L. (1922). *Tractatus-logico- philosophicus.* New York: Harcourt, Brace and Company, Inc.

Womack, J. P., Jones, D. T., & Roos, D. (1990). *The machine that changed the world*. New York: Simon & Schuster.

Woods, D. D., & Cook, R. I. (2002). Nine steps to move forward from error. *Cognition, Technology & Work, 4,* 137–144.

Woods, D. D., & Hollnagel, E. (2006). *Joint cognitive systems: Patterns in cognitive systems engineering*. Boca Raton, FL: CRC Press.

Woods, D. D., Johannesen, L. J., Cook, R. I., & Sarter, N. B. (1994). *Behind human error: Cognitive systems, computers, and hindsight* (State-of-the-art report). Columbus, OH: CSERIAC.

Woods, D. D., Licu, T., Leonhardt, J., Rayo, M., Balkin, E. A., & Ciopoena, R. (2021). *Patterns in how people think and work: Importance of patterns discovery for understanding complex adaptive systems: A white paper*. Brussels: Eurocontrol.

Zadeh, L. A. (2014). A note on modal logic and possibility theory. *Information Sciences, 279,* 908–913.

Zwetsloot, G. I., Aaltonen, M., Wybo, J. L., Saari, J., Kines, P., & De Beeck, R. O. (2013). The case for research into the zero accident vision. *Safety Science, 58,* 41–48.

Index

Page numbers in *italics* and **bold** indicate Figures and Tables, respectively.